T0269506

Gut efficiency; the key ingredient in ruminant production

Gut efficiency; the key ingredient in ruminant production

Elevating animal performance and health

edited by:

Sylvie Andrieu

David Wilde

Wageningen Academic
P u b l i s h e r s

ISBN: 978-90-8686-067-8

First published, 2008

Wageningen Academic Publishers
The Netherlands, 2008

All rights reserved.
Nothing from this publication may be
reproduced, stored in a computerised system
or published in any form or in any manner,
including electronic, mechanical, reprographic
or photographic, without prior written
permission from the publisher, Wageningen
Academic Publishers, P.O. Box 220,
NL-6700 AE Wageningen, The Netherlands.
www.WageningenAcademic.com

The individual contributions in this
publication and any liabilities arising from
them remain the responsibility of the authors.

The publisher is not responsible for possible
damages, which could be a result of content
derived from this publication.

Contents

Optimized rumen and gut health in dairy cattle: the U.S. approach

M.F. Hutjens
Department of Animal Sciences, University of Illinois, Urbana IL 61801, USA;
hutjensm@uiuc.edu

1. Introduction

Optimising rumen and lower gut health are key focus areas for U.S. dairy managers, consultants and veterinarians. A healthy rumen leads to a healthy cow with optimum milk yield, desirable milk components, optimal reproduction, and enhanced immune function. Another aspect of gut health is hemorrhagic bowel syndrome (HBS) which continues to cause significant losses on some U.S. dairy farms.

2. Carbohydrate metabolism in the rumen

Carbohydrates contribute 70 to 80 percent of the diet dry matter while protein, fat, and minerals make up the remaining portion (Hutjens, 2003). Carbohydrates are the primary energy source for the cow and support rumen function and microbial growth. Two carbohydrate categories occur in feeds: cell solubles (sugar and starch) and cell walls (cellulose, hemicellulose, lignin and pectin). Sugar, starch, and fibre are digested by rumen microbes converting carbohydrates to volatile fatty acids (VFA) (Table 1). These VFA are the main source of energy. When the VFA ratios and levels shift, milk yield and components change. Rumen availability and digestibility of cell wall and cell solubles vary depending on growth stage and maturity (forages), source of carbohydrate (starch or cellulose) and processing (grinding of grain or chopping of forages). Table 2 illustrates the effect of grain source and processing on starch digestion in the rumen. Dairy farmers and nutritionists must decide the correct source and rate of starch fermentation in the rumen based on rumen pH, forage sources, level of non-fibre carbohydrate, dry matter intake and price of starch containing grains. If more rumen fermentable carbohydrate is needed, finely processed corn, steam-flaked corn, high moisture corn, wheat or barley grain, and/or corn starch could be added. If subacute acidosis is occurring, lowering starch levels, shifting to corn gluten feed or other by-product feeds could be the correct decision.

Table 1. Characteristics of rumen bacterial groups (Hutjens, 1998).

Bacteria class	Substrate preference	Nitrogen needs	Main VFA produced	pH range	Time to double (hr)
Fibre	cellulose, hemicellulose	ammonia	acetate, butyrate	6.2-6.8	8-10
Starch and sugar	starch, sugar	ammonia, amino acids	propionate, lactate	5.5-6.0	1-2

Table 2. Percent starch fermented in the rumen based on grain type and form (Hutjens, 1998).

Grain	Ground % of starch	Whole % of starch
Oats	94	59
Wheat	93	78
Corn (steam flaked)	86	na
Barley	78	65
Sorghum	78	65
Corn	72	61

2.1. Volatile fatty acid production

End products of microbial digestion are VFA which are absorbed from the rumen and serve as a source of energy for the dairy cow (NRC, 2001). The primary VFA is acetate which is a two carbon VFA. This represents 55 to 70 percent of the total VFA production and produced from the digestion of fibre (Table 1). Propionate or propionic acid is a three carbon VFA produced by starch and sugar digestiing bacteria. Propionate is converted to glucose by the liver. Glucose is used to synthesise milk lactose sparing amino acids from gluconeogenesis. The level of propionate varies from 15 to 30 percent of the total VFA production. The third main VFA is butyrate and contributes 5 to 15 percent of the VFA produced. When evaluating VFA patterns, the ratio of acetate (A) to propionate (P) reflects the rumen fermentation pattern. Under optimal rumen fermentation conditions, the A:P ratio should be greater than 2.2 to 1 (Shaver, 2005). High levels of acetate can indicate a high fibre-low fermentable carbohydrate ration. High levels of propionic acid can indicate reduced fibre digestion and acidosis. VFA analysis in the field in not available but would be a useful tool to evaluate rumen fermentation and digestion.

2.2. Rumen pH effects

Growth of fibre digesting bacteria growth is favoured within a pH range from 6.0 to 6.8 whilst starch digesting bacteria growth is favoured by a pH from 5.5 to 6. High producing cows must maintain a pH near 6.0 for optimal growth of both bacteria populations resulting in a favourable VFA pattern and yield. The type of diet can shift pH with high forage rations favouring a pH over 6. Forages stimulate higher rates of saliva secretion which contains bicarbonate that buffers the rumen and increases acetate production. The carbohydrate in forage (cellulose and hemicellulose) is not degraded as rapidly by the rumen microbes as are carbohydrates in concentrates (starch and sugar). Legume forages also have a higher natural buffering capacity.

The physical form of feeds (grinding, pelleting, and chopping) will change the size of the feed particle. If forage particle size is too short, a forage mat in the rumen will not be maintained, fibre digestion will be decreased and rumen pH lowered. Saliva production is also reduced due to less cud chewing time. Cows will typically spend over 500 minutes of chewing time per day, 26 to 33 minutes of chew time per kilogram of dry matter and 60 to 70 percent of the cows should be chewing their cuds when resting. If concentrates are ground too fine, starch is exposed to microbial digestion and increased degradation. Rumen pH drops and propionic acid production increases changing milk components (less milk fat percentage and higher milk protein percentage) and lower milk yield. Steam flaking, pelleting or grinding will change starch structure (more available in the rumen for fermentation) which can be beneficial (increases rumen microbial growth) or negative (increases the risk of rumen acidosis).

The level of feed intake changes rumen degradation and synthesis. Rumen pH can drop as more substrate (such as starch) is available for microbial use increasing acid production (negative effect). The amount of saliva produced per unit of dry matter can also decline. Wet rations can reduce rumen pH due to less saliva production to wet the feed for swallowing. If the wet feed is silage, less chewing is needed to reduce particle size lowering rumination time. Silage can have a pH below 4 increasing acid load. Adding sodium bicarbonate to corn silage raising pH above 5 may increase intake prior to feeding. If the total ration dry matter exceeds 55 percent due to ensiled and fermented feeds, dry matter intake can be reduced.

Adding unsaturated fats and oils (such as vegetable and fish oils) can reduce rumen pH and shift VFA patterns. Unsaturated fatty acids can reduce fibre digestibility, decrease rumen pH, be toxic to fibre digesting bacteria, and/or coat fibre particles reducing fibre digestion. Processing of oilseeds (such as grinding or extruding) can rupture the cell wall of the seed and release the oil in the rumen. Feeding whole oil seeds can reduce this risk. Limit oil from oilseeds to 0.45 to 0.70 kilogram per cow per day (0.22 kg if the oil was extracted and fed as "free" oil).

The method of feeding will change the rumen environment. TMR (total mixed rations) stabilises rumen pH, synchronises DIP and fermentable carbohydrate, increases dry matter intake and minimises feed selection. If concentrates are fed separately, limit the amount to 3 kg DM per meal, avoid high levels of starch-containing grains, and evaluate the effect of feed processing.

Rumenocentesis (rumen tap) is a field technique to determine rumen pH and VFA concentrations from intact cows (Nordlund and Garrett, 1994). A 13 cm, 16 gauge needle is inserted through the rumen wall into the ventral rumen and rumen fluid is aspirated. Wisconsin researchers suggest six cows per group (fresh cows and high producers). Samples should be taken 2 to 4 hours after feeding to measure the lower values in the rumen. The pH is measured immediately after collecting. Cow testing above 5.9 are classified as normal while cows below 5.5 are considered abnormal. Evaluate the cows that are in the abnormal range, not the average value. Several factors impact changes in rumen pH.

2.3. Fibre relationships

Acid detergent fibre (ADF) consists of cellulose, lignin, lignified nitrogen compounds (such as heat damaged proteins) and insoluble ash. Forage laboratories use ADF to predict energy concentration or digestibility. Neutral detergent fibre (NDF) is becoming the fibre analysis of choice and consists of ADF plus hemicellulose (total cell wall content). NDF is correlated to feed intake and chewing time. Forage NDF refers to the percent or amount of NDF in a dairy ration based only on forage sources (hay, silage, and fresh forage). It is used as an index of rumination and forage mat formation in the rumen. No adjustments are made for length of particle size and type of forage. Physically effective NDF refers to the proportion of NDF from all feeds (forage and concentrates) that contribute to physical fibre. Each feed's particle size (based on screen separations) is assigned an effective NDF percentage which is multiplied by the level of NDF and amount of dry matter fed. Coarse chopped hay silages have higher values (70 to 80 percent physically effective NDF) while finely chopped hay silage could be as low as 25 percent.

Two fibre requirements are needed for optimum rumen function: chemical fibre concentration (measured as the percent ADF and NDF in the total ration dry matter) and fibre length (measured as physically effective NDF or forage NDF). Commercial labs can measure forage particle size by screening silage and TMR samples. Penn State developed a simple feed particle separator using three or four boxes with hand shaking. The amount in each box can be plotted to determine if particle size meet minimal needs. Physical form must be evaluated as forage harvester can chop more precisely, silo unloaders can shorten forage length, and TMR mixers can reduce particle size.

2.4. Subacute rumen acidosis (SARA)

Rumen acidosis is the number one metabolic disorder diagnosed by the University of Wisconsin Veterinary College (Krause and Oetzel, 2006). Two types of acidosis are reported in the field: acute and subacute acidosis. Acute acidosis is less common and severe. Affected animals are depressed, off-feed, elevated heart rate, diarrhoea and may die. Cows experiencing subacute rumen acidosis have mild diarrhoea, lower dry matter and haemorrhages in the hoof. Rumen pH drops below 6 and remains low for several hours. Further, VFA patterns shift (higher levels of propionate with an acetate to propionate ratio <2.2). Diagnosing subclinical acidosis in the field is a challenge. The following signs can be useful but can vary and be caused by other factors (Stone, 2004):

- Cows experiencing laminitis and foot problems, especially first lactation and fresh cows.
- Cows fed more than 3 kg of concentrate dry matter per meal.
- Increasing concentrate intake after calving faster than 0.75 kg per day.
- Shifting dry cows to the high group TMR after calving without a transition ration.
- Individual cow's one full fat test point below the herd average (for example. cows below 2.6 when the herd averages 3.6 percent milk fat).
- Individual cows have milk protein tests >0.4 percentage point higher than milk fat test (for example, a cow with a 2.7% milk fat test and a 3.2% milk protein test).

- Cow craving or selectively consuming coarse long forage (straw or grass hay).
- Cows consuming sodium bicarbonate free choice (over 50 grams per cow per day)
- Manure appears loose or watery.
- Hoof surfaces have ridges or lines (hardship grooves) and/or abnormal hoof growth.
- Less than half of the cows are chewing their cud when at rest.

Wisconsin workers describe a further two types of subclinical acidosis (Krause and Oetzel, 2006): fresh cow acidosis occurs 7 days before calving to 20 days postpartum and is related to a lack of a transition diet or management factors at calving. These cows are at risk because the rumen papillae need time to elongate for optimum VFA absorption, rumen microbes must to shift to digest high energy rations and dry matter intakes slowly increase. Adapted acidosis affects cows 40 to 150 days in milk or longer. Rumen adaptation should have occurred and these cows were receiving diets that are short in functional fibre, high in starch and/or the feeding system allows for feed selection. Both types of acidosis can be occurring and require different strategies to correct.

2.5. Managing rumen health

Balancing rations for optimal microbial growth will allow for performance and health (Beauchemin *et al.*, 2006). Protein levels include 28 to 33% soluble protein of total protein, 33 to 36% RUP (rumen undegraded protein) of total protein and use of a rumen model program to meet MP (metabolisable protein) recommendations. Carbohydrates in feeds (Table 3) include different sources and type as sources of rumen fermentable carbohydrate including 22 to 26 percent starch, 4 to 6 percent sugar, 6 to 8 percent soluble fibre, 28 to 33% NDF (neutral detergent fibre), 18 to 21% ADF (acid detergent fibre) and 3 to 4% lignin. Fats/oils must be carefully monitored. Lipids can negatively impact the rumen environment. Suggested levels of lipid types include less than 5.5% total fat/oil, a maximum of 225 gram of free oil (not in seed form such as distillers grains and extruded soybeans), 454 grams of oil in seed form (soy, sunflower, canola, and/or cotton) and use of rumen inert fat/oil sources over 5.5% fat in the ration dry matter (Hutjens, 2003).

Sorting in the feed bunk can lead to SARA (Shaver, 2005). Monitoring cow behaviour including pushing and shifting feed with the muzzle, the appearance of holes in the feed bunk, the lack of aggressive eating behaviour and appearance

Table 3. Comparison of rumen fermentable carbohydrates of various feed ingredients (Hutjens, 2003).

Feed ingredient	% Starch	% Sugar	% Soluble fibre
Wheat grain	64	2	3
Barley grain	58	2	3
Bakery waste	45	8	2
Corn distiller grain	3	4	8
Corn gluten feed	20	2	3
Hominy	49	4	2
Wheat midds	22	5	6
Molasses	0	61	0
Whey	0	69	0
Beet pulp	1	8-20	21
Citrus pulp	2	24	34

of longer feed particles (change in five percent of original TMR compared to the remaining feed in the bunk using the Penn State Particle Box). To reduce sorting, reduce feed (forage) particle below 5 cm to lower the ability of the cow to separate longer pieces, add water (3 litres per cow) or molasses based-liquid supplement, increase the number of feedings and push-up of remaining TMR more frequently. Feed bunk management principles that can reduce sorting risks are: monitor feed refusal levels (2 to 4% weigh-backs); provide 60 cm of bunk space per cow (75 cm for fresh cows); provide a smooth feed bunk surface (concrete finish, plastic liner, and/or epoxy sealant).

Feed additives can optimise rumen performance (Hutjens, 2003). Sodium bicarbonate (buffer) can be effective in rations that are high in wet silages, component fed herds, moderate to high starch diets, marginal in forage particle length and low in saliva production (cud-chewing activity). The recommended level of sodium bicarbonate or sodium sesquicarbonate is 0.75 percent of the total ration dry matter (200 to 250 grams per cow per day). Do not depend on free-choice buffer consumption – force-feed the recommended amount if needed. Yeast culture products can increase dry matter intake, reduce lactic acid levels and improve the rumen environment to increase VFA yield. The recommended level of yeast culture product will vary and should be fed according to manufacturer guidelines based on research results. Mycotoxin sequestering agents continue to grow in their use as more ensiled feeds are fed and secondary growth on silage faces, or in feed bunks, can lower feed intake – increasing the risk of mycotoxin formation and negatively reducing rumen microbial yield. Direct fed microbial products will require additional research to determine the optimal level of activity and species of microbes to predict when an economic response will occur.

Several approaches to evaluating rumen function on the farm can be used (Hall, 2006). Faecal consistency reflects intake and rate of passage of the ration. A manure score between 2.75 to 3.5 (using a five point scale with 1 equal to watery and 5 stiff) is desirable. Scoring 20 to 30 cows per group (early lactation, late lactation, being first lactation cow or other groups) can result in a useful average to monitor. Less than 10 percent of lactating cows should be one full milk fat test point below herd or breed average (3.7% vs. 2.7% for example). MUN (milk urea nitrogen) values over 14-16 mg/dl can indicate excessive protein intake, poor capture of rumen ammonia, low fermentable carbohydrate and/or SARA. Lameness herd scores averaging less than 1.4 (on a five point basis with 1 showing no evidence of locomotion problem to a score 5 which is a cow severely lame) are recommended. A cow comfort index greater than 85% (percent of cows lying down, eating or drinking divided by total number of cows) can be important when interpreting lameness scores (Garrett *et al.*, 1999).

3. Lower gut function

Haemorrhagic bowel syndrome (HBS) is a deadly lower digestive tract disorder that is increasing in frequency in adult U.S. dairy cows (Kirkpatrick *et al.*, 2001). It is characterised by a rapid influx of blood in to the small intestine of dairy cattle. forming clots within the intestine creating an obstruction. Dairy cattle usually die within 48 hours after clinical signs with fatality rates over 85 percent. The cause of HBS has not been clearly defined but *Clostridium perfringens* type A and/or the presence of a common mould (*Aspergillus fumigatus*) has been implicated (Forsberg and Wang, 2006). Nutritional contributing factors include high level of fermentable carbohydrates, feeding of TMR related to sorting risk and/or presence of the organism(s) listed above in the feed or lower gut.

Possible risk factors based on a Minnesota field survey and national dairy survey in 2002 included the following (Godden *et al.*, 2001):
• Parity: more common in second and older lactation cows.
• Herd size: more common on large dairy farms (>100 cows).
• Stage of lactation: higher occurrence in early lactation (<100 days in milk).

- Feeding system: greater frequency in herds fed a TMR.
- Level of milk production: increase in HBS as rolling herd average increased.
- Season: occurred in winter and fall months.
- Region: higher incidence in the western region of the U.S.

The fact that *C. perfringens* type A is normally found in low numbers in the lower gastro-intestinal tract of healthy cows raised concerns as to whether or not this is a causative organism, as the numbers rapidly increase after death. Research at Oregon State University (Table 4) has reported that a common mould (*A. fumigatus*) was associated with HBS that can suppress blood clotting leading to uncontrolled bleeding (Forsberg, 2003). Cows that are stressed or immuno-suppressed may have a higher risk of HBS when the common mould is consumed (Forsberg and Wang, 2006). Based on field observations and survey data, the following actions have been reported to be useful to slow or lower the rate of HBS (Aleksih, 2004):

- Lower the amount and rate of fermentable carbohydrates. Reducing the level of shelled corn and increasing baled hay are common practices.
- Vaccinating with a clostridium product has varied success.
- Use of antifungal products has been reported to be successful by Oregon workers (Forsberg and Wang, 2006).

Table 4. Detection of pathogens associated with cows dying from HBS or other GI diseases (Forsberg and Wang, 2006).

Cause	Sample (#)	C. perfringens (# postive)	Salmonella (# postive)	A. fumigatus (# positive)
HBS	16	14	1	13
Other GI tract diseases	9	6	4	0

References

Alekish, M.O., 2004. Progress in the understanding of hemorrhagic bowel syndrome. *Tri-State Dairy Nutr. Conf Proc,* pp. 37-40.

Beauchemin, K.A, W.Z. Yang and G. Penner, 2006. Ruminal acidosis in dairy cows: balancing effective fiber with starch availability. *Pacific Northwest Anim. Nutr. Conf. Proc.* pp 4-16.

Forsberg, N.E. and Y. Wang, 2006. Nutrition and immunity in dairy cattle: Implications to Hemorrhagic bowel syndrome. *Mid-South Rum. Nutr. Conf. Proc.* **11**:20.

Forsberg, N., 2003. New findings on jejunal hermorrhagic syndrome. *Hoard's Dairyman Magazine* **148**:311 (April).

Garrett,E.F. M.N. Pereira, K.V. Nordlund, L.E. Armentano, J.W. Goodger, and G.R. Oetzel, 1999. Diagnostic methods for detecting subacute ruminal acidosis in dairy cattle. *J. Dairy Sci.* **82**:1170-1178.

Godden, S., R. Frank, and T. Ames, 2001. Survey of Minnesota dairy veterinarians on the occurrence of and potential risk factors for jejunal hemorrhage syndrome in adult dairy cows. *Bovine Pract.* **35**:97-103

Hall, M.B., 2006. Rumen acidosis: carbohydrate feeding considerations. *Penn State Dairy Cattle Nut. Proc.* pp. 1-9.

Hutjens, M.F., 1998. Rumen acidosis. *http://www.livestocktrail.uiuc.edu/dairynet/ paperDisplay.cfm?ContentID=215*

Hutjens, M.F., 2003. Feeding Guide 2nd ed. *Hoard's Dairyman Publications.* Fort Atkinson, WI.

Krause, M.K. and G.R. Oetzel, 2006. Understanding and preventing subacute ruminal acidosis in dairy herds: a review. *Anim. Feed Sci. Technology* **126**:215-236.

Kirkpatrick, M.A., L.L. Timms, K.W. Kersting, and J.M. Kinyon, 2001. Case report-- jejunal hemorrhage syndrome of dairy cattle. *Bovine Pract.* **35**:104-116.

National Research Council, 2001. Nutrient requirements of dairy cattle. 7th ed rev. *Natl. Acad. Sci.* Washington, D.C.

Nordlund, K.V. and E.F. Garrett, 1994. Rumenocentesis: a technique for the diagnosis of subacute rumen acidosis in dairy herds. *Bovine Pract.* **28**:104.

Shaver, R.D., 2005. Feeding to minimize acidosis and laminitis in dairy cattle. *Cornell Nutr. Conf. Proc.* pp. 1-8.

Stone, W.C., 2004. Nutritional approaches to minimize ruminal acidosis and laminitis in dairy cattle. *J. Dairy Sci.* **87**(E Suppl): E12-26.

Digestive health and pathogenic load in cattle: impact on food safety

S. Calsamiglia
Dpt Ciencia Animal i dels Aliments, Universitat Autonoma de Barcelona, 08193 Bellaterra, Spain; Sergio.Calsamiglia@uab.es

1. Introduction

Food-born diseases are a major public health concern and strict control measures are in place to reduce its prevalence. The most prevalent food-born infections are caused by consumption of poultry products infected with *Salmonella* and *Campylobacter* (Weimer, 2000). However, a new food-born infection caused by a highly toxigenic strain of *Escherichia coli* (*E. coli O157:H7*) has been detected in beef (Riley *et al.*, 1983). Whilst the prevalence of this bacteria is 10 times lower than *Salmonella* and *Campylobacter* infections, the rate of hospitalisation is larger in *E. coli O157:H7* and caused 73,000 illnesses, accounting for about 60 to 250 deaths per year in the US (Mead *et al.*, 1999; Armstrong *et al.*, 1996).

Escherichia coli is a gram-negative bacteria that colonises the gastrointestinal tract of mammals. In normal conditions, *E. coli* is harmless to the host but some strains (as with *E. coli O157:H7*) may produce enterotoxins that can result in haemorrhagic diarrhoea. *E. coli O157:H7* was first identified in 1982 when isolated from faeces of humans after an outbreak of haemorrhagic diarrhoea (Riley *et al.*, 1983). This strain produces two toxins known as hemolysin (that produces hemolysis) and intimin (that favours the adhesion of the bacteria to the intestinal epithelium). While cattle are asymptomatic carriers, the disease may cause serious health problems and even death in humans at a very low infection dose (Mead *et al.*, 1999; Armstrong *et al.*, 1996).

Of all *E. coli O157:H7* infections, more than 75% have been associated with the consumption of beef, although cases with other meats, fruits and vegetables that have been fertilised with cattle manure have also been reported (USDA: APHIS:VS, 1997). Actual shedding of *E. coli* in faeces is sporadic and highly dependent on the season and this prevalence is strongly correlated with the incidence of the disease in humans (Figure 1). It has been reported that up to 80% of all feedlot cattle may be infected during summer months, while less than 10% in winter months (Elder *et al.*, 2000; Keen *et al.*, 1999). In the EU, average

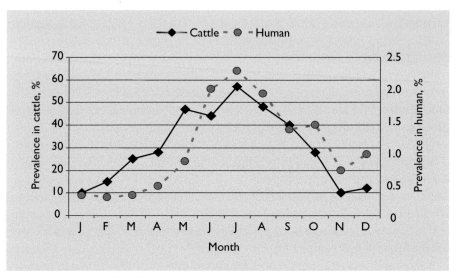

Figure 1. Changes in faecal E. coli 157:H7 shedding in cattle and its relation with human infections (Hancock et al., 1998).

prevalence may be up to 18% in cattle (Zschock *et al.*, 2000), although the low rates of diagnosis may be related to the methods used for identification.

Efforts to control and reduce the prevalence of the disease have focused on post-slaughter measures – including hygienic protocols, critical control point strategies and cooking – and have been effective in reducing carcass contamination (Buchanan and Doyle, 1997; Elder *et al.*, 2000). However, a positive animal entering the slaughterhouse is the main source of carcass contamination and the strong correlation between faecal shedding of *E. coli* and level of carcass contamination emphasises the need to reduce the load of *E. coli* at preharvest level (Elder *et al.*, 2000).

2. Preharvest control measures

Because the level of contamination of cattle entering the slaughterhouse has an exponential effect in the content of *E. coli* down the food-chain, intervention at the preharvest level should be efficient and cost-effective. In fact, Jordan *et al.* (1999a,b), using empirical and mathematical models, indicated that preharvest intervention should be effective in reducing the contamination of meat with *E. coli*. The first problem in addressing preharvest control measures is the difficulty of identifying infected animals because cattle are asymptomatic carriers and faecal shedding may be sporadic (Callaway *et al.*, 2004). Moreover, widespread distribution, high persistency in the environment and high rates of reinfection of *E. coli O157:H7* make it unlikely that strategies to eradicate the disease would be successful. Therefore, efforts should concentrate in intervention measures that result in the reduction of the load of bacteria in animals entering the slaughterhouse. These strategies can be organised in four groups: (1) feeding strategies, (2) exposure reduction strategies, (3) exclusion strategies and (4) antipathogenic strategies.

2.1. Feeding strategies

In normal conditions, the low pH and the high enzymatic activity of the stomach acts as a natural barrier for bacterial infection (Waterman and Small, 1998). However, in some conditions, an extreme acid resistant *E. coli* strain develops and seems to be responsible for *E. coli* intoxication in humans. This resistance appears when bacteria have been previously exposed to a low pH or high concentrations of VFA in the media (Buchanan and Edelson, 1996). Tkalcic *et al.* (2000) indicated that *E. coli O157:H7* incubated in rumen fluid from cattle fed a high concentrate diet were more resistant to acid shock than those incubated in rumen fluid from cattle fed a high forage diet and the authors suggested that the ability of *E. coli O157:H7* to become acid-resistant may be the determinant factor that influences shedding. Diez-Gonzalez and Russell (1999) observed that the degree of acid-resistance of *E. coli* was strongly correlated with the concentration of undissociated VFA (Figure 2), which is itself dependent upon the total concentration of VFA and the pH of the media (with acetate, propionate and butyrate the strongest inducers of extreme acid resistance). In normal conditions, the limited feed available for fermentation in the large intestine results in a lower concentration of VFA and higher pH and common strains of *E. coli* grow. However, when cattle are fed a high concentrate diet, some starch reaches the large intestine undigested, undergoing a secondary fermentation. This secondary fermentation increases the VFA concentration and reduces pH, both factors contributing to the increase in the concentration of undissociated VFA that favours the development of these extreme acid resistant *E. coli* strains (Diez-Gonzalez *et al.*, 1998). These changes in VFA concentration and pH appear to favour mostly the pathogenic *E. coli O157:H7* strains, that may increase by 2 to 3 logs, compared with the moderate growth of non-pathogenic *E. coli* strains. Russell *et al.* (2000) suggested that there were three reasons to justify the decrease in *E. coli O157:H7* load when feeding hay: (a) it forms a ruminal mat that increases retention time of grain in the rumen allowing a more extensive degradation, (b) it increases ruminal pH, allowing

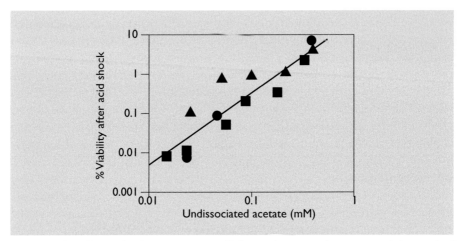

Figure 2. The relationship between extracellular undissociated acetate and the survival of wild-type Escherichia coli O157:H7 *after acid shock (pH 2, 6 h). The triangles show an experiment in which pH was 7.0 and acetate concentration changed. The squares show an experiment in which acetate concentration was 3 mM and pH changed with HCl. The circles show an experiment in which glucose concentration was varied and* E. coli 0157:H7 *produced acetate and decreased pH. Figures redrawn from the data of Diez-Gonzalez* et al. *(1998).*

a more complete fermentation and (c) it pulls buffers into the lower tract. Therefore, changes occurring in the hindgut play a major role in the shedding of toxicogenic *E. coli* strains.

Diez-Gonzalez *et al.* (1998) reported that in high grain fed cattle, *E. coli* counts in faeces increased 100-fold and strains recovered from faeces were 1000-fold more resistant to an acid shock conditions. In contrast, when cattle were shifted from a high concentrate diet into a high forage diet the concentration of *E. coli* declined 100,000-fold. Diez-Gonzalez *et al.* (1998) demonstrated that it takes about 6 days to reduce *E. coli* load after the shift from a high concentrate to a high forage diet. Keen *et al.* (1999) also observed that 53% of cattle fed a 90% concentrate diet were positive to *E. coli O157:H7* but after a change onto a 100% forage diet, only 18% were positive, suggesting that short periods of time of feeding only hay prior to slaughter reduces the load of *E. coli O157:H7* in cattle. These effects have also been reported by other authors (Jordan and McEwen, 1998; Gregory *et al.*, 2000; Scott *et al.*, 2000; Stanton and Schutz, 2000). Based on these studies, Diez-Gonzalez *et al.* (1998) suggested that cattle should be switched into a 100% hay diet 5 days prior to slaughter. Such change has minor effects on dry matter intake, body weight gain and carcass quality (Stanton and Schutz, 2000). However, the beneficial effect of a diet change on *E. coli* shedding is controversial and other authors reported no effects of hay feeding prior to slaughter (Hancock *et al.* 1998, Kudva *et al.*, 1995, 1997; Hovde *et al.*, 1999; Buchko *et al.*, 2000a,b).

Russell *et al.* (2000) also suggested that measures to control the pH in the hindgut should be tested. Although bicarbonate is commonly used as a buffer to control ruminal acidosis, there is no evidence that it reduces the pH in the large intestine or affects *E. coli* shedding. It is more likely that the more insoluble limestone and magnesium oxide may increase large intestine pH (Stokes *et al.*, 1986), although there is no evidence that it reduces *E. coli* shedding. Other feeding strategies that have been reported to affect *E. coli* shedding include barley feeding, abrupt weaning and feeding corn silage, all being associated with an increased shedding (Dargatz *et al.*, 1997; Herriot *et al.*, 1998).

2.2. Exposure reduction strategies

Cattle are exposed to three main external sources of contamination: (a) water, (b) feed and (c) dirt. Water is likely the major reservoir for reinfection of cattle and there is a strong correlation between the prevalence of *E. coli* O157:H7 in water and cattle (LeJeune *et al.*, 2004; Davis *et al.*, 2005). However, many of the common strategies to reduce bacteria in water, such as ozonation and chlorination, have had a minor impact on *E. coli* concentration in water (LeJeune *et al.*, 2004; Zhao *et al.*, 2006). Feed may also play a role as *E. coli* reservoir, where the prevalence has been estimated in the range of 1-15% of samples (Davis *et al.*, 2003 Dodd *et al.*, 2003) but a correlation between infected feed and cattle has not been yet demonstrated (Dodd *et al.*, 2003; LeJeune *et al.*, 2004). Therefore, although hygienic control of feed and water needs to be implemented in farms for many different reasons, including the control *E. coli* (Crump *et al.*, 2002), LeJeune and Wetzel (2007) suggested that the difficulties in controlling it may not always justify the cost. On the other hand, the level of dirt on animals entering the slaughterhouse is the main route of contamination of carcasses, and attempts to decrease it should be considered (Hancock *et al.*, 1998). Gregory *et al.* (2000) reported that fasted animals or animals fed hay before transport were cleaner than those fed pasture, further supporting the suggestion that feeding hay prior to slaughter may be a successful strategy to reduce *E. coli* O157:H7 load and carcass contamination.

2.3. Exclusion strategies

Competitive exclusion strategies are based on the introduction of non-pathogenic bacteria or molecules that will compete with pathogenic bacteria for nutrients or sites of attachment to the intestinal epithelia, therefore reducing the likelihood of survival of the pathogenic bacteria in the intestinal tract (Nurmi *et al.*, 1992; Steer *et al.*, 2000). Previous research has shown the effectiveness of these strategies in preventing *Salmonella* colonisation in broilers (Nurmi *et al.*, 1992), chikens (Nisbet *et al.*, 1996) and swine (Anderson *et al.*, 1999; Fedorka-Cray *et al.*, 1999). Similar strategies have also been used in preventing *E. coli* colonisation in swine (Genovese *et al.*, 2000; Harvey *et al.*, 2003). The studies using competitive exclusion strategies in cattle are more limited. Zhao *et al.* (1998) isolated several *E. coli* strains from faeces of cattle that, when inoculated two days prior to an experimental infection with *E. coli O157:H7*, resulted in a reduction in the shedding of *E. coli O157:H7* in faeces (Figure 3). The prevalence of *E. coli O157:H7* in the rumen was reduced from 26 to 14 days, whilst the prevalence in faeces was reduced from 32 to 19 days. The authors argued that such reduction was probably due to the production of inhibitory metabolites or colicins. Similar approach was also used by Zhao *et al.* (2003) and Brashears *et al.* (2003a). Brashears *et al.* (2003b) tested the effects of lactic acid bacteria on *E. coli O157:H7* shedding by cattle. Several lactic acid bacteria were isolated from faeces of cattle negative to *E. coli O157:H7* and used as probiotics. Of all strains isolated, 13 significantly reduced *E. coli* counts in faeces and 15 reduced *E. coli* counts in rumen fluid. In feedlot cattle fed a high concentrate diet, lactic acid bacteria reduced *E. coli* load by 50% (Brahears *et al.*, 2003a,b; Moxley *et al.*, 2003).

Another strategy is the use of unique molecules that may prevent the attachment of pathogenic bacteria to the intestinal wall by competing for sites of attachment. Many gastrointestinal bacteria use mannose specific lectins as receptors for the attachment to the intestinal wall. If such receptors are blocked, then the rate of colonisation and infection may be reduced. Mannan oligosaccharides have been used to interact with bacteria binding sites and reduce attachment. Salit and Gotschlich (1977) reported that *E. coli* would not attach to mammalian cells in a mannose rich media. Using a chicken model, Spring *et al.* (2000) demonstrated that mannan oligosaccharides (Bio-Mos®, Alltech Inc, USA) reduced the colonisation of *Salmonella* and *E. coli*. Although this approach appears promising, there are not yet any reports on the potential effect of mannan oligosaccharides in cattle as a method of reducing *E. coli O157:H7* shedding.

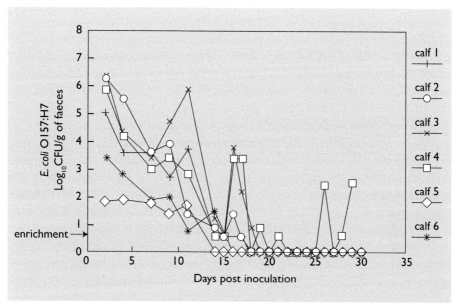

Figure 3. Shedding of E. coli *in faeces of calves administered with prebiotic bacteria and experimentally infected with* E. coli *O157:H7 two days later. Bacterial enumeration was performed by surface plating on SMA-NA plates, in duplicate. The arrow indicates that detection of* E. coli *O157:H7 was by an enrichment procedure in which 9 g of faeces was positive for* E. coli *O157:H7 (Zhao et al., 1998).*

2.4. Antipathogen strategies

An alternative is to use products or strategies that will directly reduce or eliminate the bacteria from the animal. There are three potential strategies: (1) the use of antibiotics or microbial inhibitors, (2) the use of bacteria-specific bacteriophages, and (3) the use of vaccination or antibodies.

2.4.1. Microbial inhibition

Ionophores are effective against gram-positive bacteria and, therefore, should not affect *Salmonella* or *E. coli*. However, the inhibition of gram-positive bacteria may reduce competition for nutrients or leave open niches for the development of gram-negative bacteria, resulting in an increased risk of *E. coli* growth. Edrington *et al.* (2003a,b) explored this potential risk and concluded that the use of ionophores does not increase the load of gram-negative bacteria. Other antibiotics effective against gram-negative bacteria have been tested. Elder *et al.* (2002) and Ransom *et al.* (2003) reported that the addition of neomycin significantly reduced *E. coli.* Although neomycin was very effective and has a short withdrawal period (only 24 h), the potential for an antibiotic to be approved for animal nutrition is limited by the risk of creating cross-resistance and the reduced social acceptance of the use of antibiotic in animal feeds.

Another potential antimicrobial strategy is the use chlorates. Nitrates are final acceptors of electrons in the respiratory pathway of anaerobic bacteria, and are transformed into nitrite. Chlorates are analogs of nitrate and are reduced to chlorite, which are highly toxic to bacteria. Feeding chlorates to cattle and sheep through the water resulted in a sharp reduction in the concentration and shedding of *E. coli* (Callaway *et al.*, 2002, 2003a). Because the effect is very strong and in a very short period of time, Anderson *et al.* (2000) suggested that a single dose 24 h prior to slaughter could be sufficient to reduce the *E. coli* load of cattle entering the slaughterhouse. The main limitation for the use of chlorates in cattle feeding is that it is not currently approved.

2.4.2. Bacteriophages

Bacteriophages are viruses that infect and kill bacteria. The identification, selection and production of specific bacteriophages against pathogenic bacteria may provide an alternative strategy to reduce *E. coli* shedding in cattle. Because bacteriophages can be very specific against a single strain of bacteria, its use in the prevention of *E. coli* infection has been proposed (Summers, 2001). Some preliminary work in rats, pigs, calves and sheep has been successful in controlling *E. coli* populations (Smith and Huggins, 1982, 1983; Huff *et al.*, 2002). Although some highly specific phages against *E. coli O157:H7* have been developed and successfully tested *in vitro*, their effects *in vivo* conditions have been limited (Kudva *et al.*, 1999; Bach *et al.*, 2002; Callaway *et al.*, 2003b). Smith and Huggins (1983, 1987) reported a reduction in *E. coli O157:H7* in calves supplied with bacteriophages against *E. coli.* While these results are encouraging, responses have not always been consistent (Kudva *et al.*, 1999; Bach *et al.*, 2002) and additional research is required to develop a fully effective strategy.

2.4.3. Immunisation

Several immunisation strategies against rumen and gut bacteria have been developed for ruminants. Shu *et al.* (2000) developed vaccines against bacteria involved in acidosis. A similar approach was used by Finlay *et al.* (2003), who used a vaccines against *E. coli O157:H7*. Moxley *et al.* (2003) reported that *in vivo*, this vaccine reduced shedding form 23 to 9% of all cattle tested. A different approach that proved to be effective was feeding polyclonal antibodies against specific bacteria. DiLorenzo *et al.* (2006) and Blanch *et al.* (2006) reported a reduction in *S. bovis* and *F. necrophorum* counts in the rumen after feeding polyclonal antibodies against these bacteria, both involved in the development of acidosis. No similar approaches have been tested with *E. coli* but the technology could provide a useful tool in the fight against *E. coli* in cattle.

3. Conclusion

Haemorrhagic diarrhoea caused by *E. coli O157:H7* is a major public health concern. Whilst postslaughter control measures are effective and currently implemented, the reduction in *E. coli* load at preharvest level is still necessary and likely very effective. The widespread distribution, high persistency in the environment and high rates of reinfection of *E. coli O157:H7,* make unlikely that strategies to eradicate the disease would be successful. Therefore, alternative strategies to reduce the prevalence and load need to be developed. The most promising options include a period of high forage feeding prior to slaughter, the use of lactic acid bacteria as probiotics, specific *E. coli O157:H7* bacteriophages and vaccination. However, additional research is required to identify the effectiveness of each of these alternatives in commercial conditions.

References

Anderson, R.C., L.H. Stanker, C.R. Young, S.A. Buckley, K.J. Genovese, R.B. Harvey, J.R. DeLoach, N.K. Keith and D.J. Nisbet, 1999. Effect of competitive exclusion treatment on colonization of early-weaned pigs by *Salmonella* serovar *cholerasuis*. *Swine Health Prod.* **12:**155-160.

Anderson, R.C., S.A. Buckley, L.F. Kubena, L.H. Stanker, R.B. Harvey and D.J. Nisbet, 2000. Bactericidal effect of sodium chlorate on *Escherichia coli* O157:H7 and *Salmonella typhimurium* DT104 in rumen contents in vitro. *J. Food Prot.* **63:**1038-1042.

Armstrong, G.L., J. Hollingsworth and J.G. Morris, Jr., 1996. Emerging foodborne pathogens: *Escherichia coli* O157:H7 as a model of entry of a new pathogen into the food supply of the developed world. *Epidemiol. Rev.* **18:**29-51.

Bach, S.J., T.A. McAllister, D.M. Veira, V.P. Gannon and R.A. Holley, 2002. Evaluation of bacteriophage DC22 for control of *Escherichia coli* O157:H7. *J. Anim. Sci.* **80**(Suppl. 1):263 (Abstract).

Blanch, M., S. Calsamiglia, N. DiLorenzo and A. DiCostanzo, 2006. Effects of feeding a polyclonal antibody preparation against *Streptococcus bovis* on rumen fermentation of heifers switched from a high forage to a high concentrate diet. *J. Dairy Sci.* **89**(Suppl. 1):132.

Brashears, M.M., D. Jaroni and J. Trimble, 2003b. Isolation, selection, and characterization of lactic acid bacteria for a competitive exclusion product to reduce shedding of *Escherichia coli* O157:H7 in cattle. *J. Food Prot.* **66:**355-363.

Brashears, M.M., M.L. Galyean, G.H. Loneragan, J.E. Mann and K. Killinger-Mann, 2003a. Prevalence of *Escherichia coli* O157:H7 and performance by beef feedlot cattle given *lactobacillus* direct-fed microbials. *J. Food Prot.* **66:**748-754.

Buchanan, R.L. and M.P. Doyle, 1997. Foodborne disease significance of *Escherichia coli* O157:H7 and other enterohemorrhagic *E. coli. Food Technol.* **51:**69-76.

Buchanan, R.L. and S.G. Edelson, 1996. Culturing enterohemorrhagic *Escherichia coli* in the presence and absence of glucose as a simple means of evaluating the acid tolerance of stationaryphase cells. *Appl. Environ. Microbiol.* **62:**4009-4013.

Buchko, S.J., R.A. Holley, W.O. Olson, V.P.J. Gannon and D.M. Veira, 2000a. The effect of different grain diets on fecal shedding of *Escherichia coli* O157:H7 by steers. *J. Food Prot.* **63:**1467-1474.

Buchko, S.J., R.A. Holley, W.O. Olson, V.P.J. Gannon and D.M. Veira, 2000b. The effect of fasting and diet on fecal shedding of *Escherichia coli* O157:H7 by cattle. *Can. J. Anim. Sci.* **80:**741-744.

Callaway, T.R., R.C. Anderson, T.S.Edrington, K.J. Genovese, K.M. Bischoff, T.L. Poole, Y.S. Jung, R.B. Harvey and D.J. Nisbet, 2004. What are we doing about *Escherichia coli* O157:H7 in cattle? *J. Anim. Sci.* **82** E-Suppl:E93-99.

Callaway, T.R., R.C. Anderson, K.J. Genovese, T.L. Poole, T.J. Anderson, J.A. Byrd, L.F. Kubena and D.J. Nisbet, 2002. Sodium chlorate supplementation reduces *E. coli* O157:H7 populations in cattle. *J. Anim. Sci.* **80:**1683-1689.

Callaway, T.R., T.S. Edrington, R.C. Anderson, K.J. Genovese, T.L. Poole, R.O. Elder, J.A. Byrd, K.M. Bischoff and D.J. Nisbet, 2003a. *Escherichia coli* O157:H7 populations in sheep can be reduced by chlorate supplementation. *J. Food Prot.* **66:**194-199.

Callaway, T.R., T.S. Edrington, R.C. Anderson, Y.S. Jung, K.J. Genovese, R.O. Elder and D.J. Nisbet, 2003b. Isolation of naturally-occurring bacteriophage from sheep that reduce populations of *E. coli* O157:H7 in vitro and in vivo. In: *Proc. 5th Int. Symp. on Shiga Toxin-Producing Escherichia coli Infections*, Edinburgh, U.K., p. 25.

Crump, J.A., P.M. Griffin and F.J. Angulo, 2002. Bacterial contamination of animal feed and its relationship to human foodborne illness. *Clin. Infect. Dis.* **35:**859-865.

Dargatz, D.A., S.J. Wells, L.A. Thomas, D.D. Hancock and L.P. Garber, 1997. Factors associated with the presence of *Escherichia coli* O157 in feces of feedlot cattle. *J. Food Prot.* **60:**466-470.

Davis, M.A., K.A. Cloud-Hansen, J. Carpenter and C.J. Hovde, 2005. *Escherichia coli* O157:H7 in environments of culture-positive cattle. *Appl. Environ. Microbiol.* **71:**6816-6822.

Davis, M.A., D.D. Hancock, D.H. Rice, D.R. Call, R. DiGiacomo, M. Samadpour and T.E. Besser, 2003. Feedstuffs as a vehicle of cattle exposure to *Escherichia coli* O157:H7 and *Salmonella enterica. Vet. Microbiol.* **95:**199-210.

Diez-Gonzalez, F. and J.B. Russell, 1999. Factors affecting the extreme acid resistance of *Escherichia coli* O157:H7. *Food Microbiol.* **16:**367-374.

Diez-Gonzalez, F., T.R. Callaway, M.G. Kizoulis and J.B. Russell, 1998. Grain feeding and the dissemination of acid-resistant *Escherichia coli* from cattle. *Science* **281:**1666-1668.

DiLorenzo, N., F. Diez-Gonzalez and A. DiCostanzo, 2006. Effects of feeding polyclonal antibody preparations on ruminal bacterial populations and ruminal pH of steers fed high-grain diets. *J. Anim Sci.* **84:**2178-2185.

Dodd, C.C., M. W. Sanderson, J.M. Sargeant, T.G. Nagaraja, R.D. Oberst, R.A. Smith and D.D. Griffin, 2003. Prevalence of *Escherichia coli* O157 in cattle feeds in Midwestern feedlots. *Appl. Environ. Microbiol.* **69:**5243-5247.

Edrington, T.S., T.R. Callaway, K.M. Bischoff, K.J. Genovese, R.O. Elder, R.C. Anderson and D.J. Nisbet, 2003a. Effect of feeding the ionophores monensin and laidlomycin propionate and the antimicrobial bambermycin to sheep experimentally infected with *E. coli* O157:H7 and *Salmonella Typhimurium. J. Anim. Sci.* **81:**553-560.

Edrington, T.S., T.R. Callaway, P.D. Varey, Y.S. Jung, K.M. Bischoff, R.O. Elder, R.C. Anderson, E.Kutter, A.D. Brabban and D.J. Nisbet, 2003b. Effects of the antibiotic ionophores monensin, lasalocid, laidlomycin propionate and bambermycin on *Salmonella* and *E. coli* O157:H7 in vitro. *J. Appl. Microbiol.* **94:**207-213.

Elder, R.O., J.E. Keen, G.R. Siragusa, G.A. Barkocy-Gallagher, M. Koohmaraie and W.W. Lagreid, 2000. Correlation of enterohemorrhagic *Escherichia coli* O157 prevalence in feces, hides, and carcasses of beef cattle during processing. *Proc. Natl. Acad. Sci. USA* **97:**2999-3003.

Elder, R.O., J.E. Keen, T.E. Wittum, T.R. Callaway, T.S. Edrington, R.C. Anderson and D.J. Nisbet, 2002. Intervention to reduce fecal shedding of enterohemorrhagic *Escherichia coli* O157:H7 in naturally infected cattle using neomycin sulfate. *J. Anim. Sci.* **80**(Suppl. 1):15 (Abstract).

Fedorka-Cray, P.J., J.S. Bailey, N.J. Stern, N.A. Cox, S.R. Ladely and M. Musgrove, 1999. Mucosal competitive exclusion to reduce *Salmonella* in swine. *J. Food Prot.* **62:**1376-1380.

Finlay, B., 2003. Pathogenic *Escherichia coli*: From molecules to vaccine. In: *Proc. 5th Int. Symp. on Shiga Toxin-Producing Escherichia coli Infections*, Edinburgh, U.K., p. 23.

Fuller, R., 1989. Probiotics in man and animals. J. Appl. Bacteriol. 66:365-378.

Genovese, K.J., R.C. Anderson, R.B. Harvey and D.J. Nisbet, 2000. Competitive exclusion treatment reduces the mortality and fecal shedding associated with enterotoxigenic *Escherichia coli* infection in nursery-raised pigs. *Can. J. Vet. Res.* **64:**204-207.

Gregory, N.G., L.H. Jacobson, T.A. Nagle, R.W. Muirhead and G.J. Leroux, 2000. Effect of preslaughter feeding system on weight loss, gut bacteria, and the physico-chemical properties of digesta in cattle. *N. Z. J. Agric. Res.* **43:**351-361.

Hancock, D.D., T.E. Besser and D.H. Rice, 1998. Ecology of *Escherichia coli* O157:H7 in cattle and impact of management practices. In: *Escherichia coli O157-H7 and Other Shiga Toxin-Producing E. coli Strains*. J.B. Kaper and A.D. O'Brien, eds. American Society for Microbiology, Washington, DC., pp. 85-91.

Harvey, R.B., R.C. Ebert, C.S. Schmitt, K. Andrews, K.J. Genovese, R.C. Anderson, H.M. Scott, T.R. Callaway and D.J. Nisbet, 2003. Use of a porcine-derived, defined culture of commensal bacteria as an alternative to antibiotics used to control *E. Coli* disease in weaned pigs. In: *9th Int. Symp. Dig. Physiol. in Pigs*, Banff, AB, Canada, pp. 72-74.

Herriott, D.E., D.D. Hancock, E.D. Ebel, L.V. Carpenter, D.H. Rice and T.E. Besser, 1998. Association of herd management factors with colonization of dairy cattle by Shiga toxin-positive *Escherichia coli* O157. *J. Food Prot.* **61**:802-807.

Hovde, C.J., P.R. Austin, K.A. Cloud, C.J. Williams and C.W. Hunt, 1999. Effect of cattle diet on *Escherichia coli* O157:H7 acid resistance. *Appl. Environ. Microbiol.* **65**:3233-3235.

Huff, W.E., G.R. Huff, N.C. Rath, J.M. Balog, H. Xie, P.A. Moore and A.M. Donoghue, 2002. Prevention of *Escherichia coli* respiratory infection in broiler chickens with bacteriophage (spr02). *J. Poult. Sci.* **81**:437-441.

Jordan, D. and S.A. McEwen, 1998. Effect of duration of fasting and short term high-roughage ration on the induction of *Escherichia coli* biotype 1 in cattle feaces. *J. Food Prot.* **61**:531-534.

Jordan, D., S.A. McEwen, A.M. Lammerding, W.B. McNab and J.B. Wilson, 1999a. Pre-slaughter control of *Escherichia coli* O157 in beef cattle: A simulation study. *Prev. Vet. Med.* **41**:55-74.

Jordan, D., S.A. McEwen, A.M. Lammerding, W.B. McNab and J.B. Wilson, 1999b. A simulation model for studying the role of pre-slaughter factors on the exposure of beef carcasses to human microbial hazards. *Prev. Vet. Med.* **41**:37-54.

Keen, J.E., G.A. Uhlich and R.O. Elder, 1999. Effects of hayand grain-based diets on fecal shedding in naturally-acquired enterohemorrhagic *E. coli* (EHEC) O157 in beef feedlot cattle. In: *80th Conf. Res. Workers in Anim. Dis.*, Chicago, IL. (Abstract).

Kudva, I.T., P.G. Hatfield and C.J. Hovde, 1995. Effect of diet on the shedding of *Escherichia coli* O157:H7 in a sheep model. *Appl. Environ. Microbiol.* **61**:1363-1370.

Kudva, I.T., C.W. Hunt, C.J. Williams, U.M. Nance and C.J. Hovde, 1997. Evaluation of dietary influences on *Escherichia coli* O157:H7 shedding by sheep. *Appl. Environ. Microbiol.* **63**:3878-3886.

Kudva, I.T., S. Jelacic, P.I. Tarr, P. Youderian and C.J. Hovde, 1999. Biocontrol of *Escherichia coli* O157 with O157-specific bacteriophages. *Appl. Environ. Microbiol.* **65**:3767-3773.

LeJeune, J.T. and A.N. Wetzel, 2007. Preharvest control of *Escherichia coli* O157 in cattle. *J. Anim. Sci.* **85**(E. Suppl.):E73-E80.

LeJeune, J.T., T.E. Besser, D.H. Rice, J.L. Berg, R.P. Stilborn and D.D. Hancock, 2004. Longitudinal study of fecal shedding of *Escherichia coli* O157:H7 in feedlot cattle: Predominance and persistence of specific clonal types despite massive cattle population turnover. *Appl. Environ. Microbiol.* **70:**377-384.

Mead, P.S., L. Slutsker, V. Dietz, L.F. McCraig, J.S. Bresee, C. Shapiro, P.M. Griffin and R.V. Tauxe, 1999. Food-related illness and death in the United States. *Emerg. Infect. Dis.* **5:**607-625.

Moxley, R.A., D. Smith, T.J. Klopfenstein, G. Erickson, J. Folmer, C. Macken, S. Hinkley, A. Potter and B. Finlay, 2003. Vaccination and feeding a competitive exclusion product as intervention strategies to reduce the prevalence of *Escherichia coli* O157: H7 in feedlot cattle. In: *Proc. 5th Int. Symp. on Shiga Toxin- Producing Escherichia coli Infections*, Edinburgh, U.K., p. 23.

Nisbet, D.J., D.E. Corrier, S. Ricke, M.E. Hume, J.A. Byrd and J.R. DeLoach, 1996. Maintenance of the biological efficacy in chicks of a cecal competitive-exclusion culture against *Salmonella* by continuous-flow fermentation. *J. Food Prot.* **59:**1279-1283.

Nurmi, E., L. Nuotio and C. Schncitz, 1992. The competitive exclusion concept: Development and future. *Int. J. Food Microbiol.* **15:**237-240.

Ransom, J.R., K.E. Belk, J.N. Sofos, J.A. Scanga, M.L. Rossman, G.C. Smith and J.D. Tatum, 2003. *Investigation of on-farm management practices as pre-harvest beef microbiological interventions.* Natl. Cattlemen's Beef Assoc. Res. Fact Sheet, Centennial, CO.

Riley, L.W., R.S. Remis, S.D. Helgerson, H.B. McGee, J.G. Wells, B.R. Davis, R.J. Hebert, E.S. Olcott, L.M. Johnson, N.T. Hargrett, P.A. Blake and M.L. Cohen, 1983. Hemorrhagic colitis associated with a rare *Escherichia coli* serotype. *N. Engl. J. Med.* **308:**681-685.

Russell, J.B., F. Diez-Gonzalez and G.N. Jarvis, 2000. Invited Review: Effects of Diet Shifts on *Escherichia coli* in Cattle. *J. Dairy Sci.* **83:**863-873.

Salit, I.E. and E.C. Gotschlich, 1977. Hemmagglutination by purified type 1 *Escherichia coli* pili. *J. Exp. Med.* **146:**1169-1181.

Scott, T., C. Wilson, D. Bailey, T. Klopfstein, T. Milton, R. Moxley, D. Smith, J. Gray and L. Hungerford, 2000. Influence of diet on total and acid resistant *E. coli* and colonic pH. *Nebraska Beef Rep.* 34-41.

Shu, Q., H.S. Gill, R.A. Leng and J.B. Rowe, 2000. Immunization with a *Streptococcus bovis* vaccine administered by different routes against lactic acidosis in sheep. *Veterinary Journal* **159:**262-269.

Smith, H.W. and R.B. Huggins, 1982. Successful treatment of experimental *E. coli* infections in mice using phage: Its general superiority over antibiotics. *J. Gen. Microbiol.* **128:**307-318.

Smith, H.W. and R.B. Huggins, 1983. Effectiveness of phages in treating experimental *Escherichia coli* diarrhoea in calves, piglets and lambs. *J. Gen. Microbiol.* **129:**2659-2675.

Smith, H.W. and R.B. Huggins, 1987. The control of experimental *E. coli* diarrhea in calves by means of bacteriophage. *J. Gen. Microbiol.* **133**:1111-1126.

Spring, P., C. Wenk, K.A. Dawson and K.E. Newman, 2000. Effect of mannan oligosaccharide on different cecal parameters and on cecal concentration on enteric bacteria in challenged broiler chicks. *Poult. Sci.* **79**:205-211.

Stanton, T.L. and D. Schutz, 2000. Effect of switching from high grain to hay five days prior to slaughter on finishing cattle performance. Colorado State Univ. Research Report. Ft. Collins, CO.

Steer, T., H. Carpenter, K. Tuohy and G.R. Gibson, 2000. Perspectives on the role of the human gut microbiota and its modulation by pro and prebiotics. *Nutr. Res. Rev.* **13**:229-254.

Stokes, M.R., L.L. Vandemarck and L.S. Bull, 1986. Effects of sodium bicarbonate, magnesium oxide and a commercial buffer mixture in early lactation cows fed hay crop silage. *J. Dairy Sci.* **69**:1595-1603.

Summers, W.C., 2001. Bacteriophage therapy. *Ann. Rev. Microbiol.* **55**:437-451.

Tkalcic, S., C.A. Brown B.G. Harmon, A.V. Jain, E.P.O. Mueller, A. Parks, K.L. Jacobsen, S.A. Martin, T. Zhao and M.P. Doyle, 2000. Effects of diet on rumen proliferation and faecal shedding of *Escherichia coli* O157:H7 in calves. *J. Food Prot.* **63**:1630-1636.

USDA:APHIS:VS, 1997. An Update: *Escherichia coli* O157:H76 in humans and cattle. Centers for Epidemiology and AnimalHealth, Fort Collins, CO.

Waterman, S.R. and P.L.C. Small. 1998. Acid-sensitive enteric pathogens are protected from killing under extremely acidic conditions of pH 2.5 when they are inoculated onto certain food sources. *Appl. Environ. Microbiol.* **64**:3882-3886.

Weimer, B., 2000. Foodborne Illness - The Preventable Diseases. In: *Proc. Intermountain Nutrition Conference.*

Zhao, T., M.P. Doyle, B.G. Harmon, C.A. Brown, P.O.E. Mueller and A.H. Parks, 1998. Reduction of carriage of enterohemorrhagic *Escherichia coli* O157:H7 in cattle by inoculation with probiotic bacteria. *J. Clin. Microbiol.* **36**:641-647.

Zhao, T., S. Tkalcic, M.P. Doyle, B.G. Harmon, C.A. Brown and P. Zhao, 2003. Pathogenicity of enterohemorrhagic *Escherichia coli* in neonatal calves and evaluation of fecal shedding by treatment with probiotic *Escherichia coli*. *J. Food Prot.* **66**:924-930.

Zhao, S., P.F. McDermott, S. Friedman, J. Abbott, S. Ayers, A. Glenn, E. Hall-Robinson, S. K. Hubert, H. Harbottle, R.D. Walker, T.M. Chiller and D.G. White, 2006. Antimicrobial resistance and genetic relatedness among *Salmonella* from retail foods of animal origin: NARMS retail meat surveillance. *Foodborne Pathog. Dis.* **3**:106-117.

Zschock, M., H.P. Hamann, B. Kloppert and W. Wolter, 2000. Shiga-toxin-producing *Escherichia coli* in faeces of healthy dairy cows, sheep and goats: prevalence and virulence properties. *Lett. Appl. Microbiol.* **31**:203:208.

Enteric methane production by ruminants and its control

J.-P. Jouany
INRA, UR1213 Herbivores, Theix, 63122 Saint Genes Champanelle, France;
jouany@clermont.inra.fr

1. Introduction

Anaerobic fermentations occurring in the rumen result in the production of methane, which is then released into the atmosphere. This phenomenon induces animal energy losses of 3% to 14% of digestible energy (DE). Furthermore, due to its high global warming power, which is 23 times that of carbon dioxide, enteric methane accounts for 2–3% of global warming.

This paper describes the major metabolic pathways involved in methane synthesis in the rumen and of the main competitive pathways for hydrogen use. It goes on to describe the dietary and animal effects acting on methane emissions by ruminants and concluding with a discussion on the various options for methane mitigation and their limits.

2. Origin of methane

Methane is a natural end-product of rumen fermentation that plays a fundamental role in the efficacy of feed digestion by rumen microbes. ATP, which is necessary for the growth of microorganisms, is primarily produced during oxidative reactions. In chemical term, an oxidative reaction corresponds to an electron removal from an atom or a molecule. In aerobic ecosystems, oxygen is used as major electron sink to generate ATP. In anaerobic, i.e., oxygen-free ecosystems, oxidative reactions result in the removal of hydrogen (H = metabolic hydrogen) from substrates or metabolites which is then used to reduce oxidized co-factors and produce reduced co-factors (see Equation 1).

$$NAD^+ \text{ (oxidized co-factor)} + H + e^- \leftrightarrow NADH \text{ (reduced co-factor)} \qquad (1)$$

However, NADH has to be re-oxidized to NAD^+ to allow the fermentation process to continue. Thus, H can be considered as an electron acceptor to give hydrogen gas (see Equation 2). As a matter of fact, many rumen microorganisms in pure culture produce hydrogen gas as one of the major end-products of fermentation.

$$2H^{(+)} + 2e^- \leftrightarrow H_2 \qquad\qquad (2)$$

Although H_2 is formed in large amounts in the rumen, it does not accumulate because it is immediately used by methanogens for reactions that generate energy for growth.

Thus, in the complex microbial ecosystem present in the rumen, fermentative bacteria produce H_2 which is then mainly eliminated by H_2-using methanogens. This metabolic chain of hydrogen, which is called *"interspecies hydrogen transfer"*, plays a central role in the energy exchanges inside the microbial ecosystem, since NADH re-oxidation can only occur when the ruminal H_2 concentration is kept below one kPa (Miller, 1995).

All CH_4-forming bacteria in the rumen belong to the *Archaea* family and to the genus *Methanobrevibacter* (Miller, 1995). They use H_2 to reduce CO_2 into CH_4 according to Equation 3. The energy supplied by the reaction is efficiently used by *Archaea* for growth.

$$CO_2 + 4H_2 \rightarrow CH_4 + 2H_2O \qquad \Delta G_0' = -135.6 kJ \qquad (3)$$

Formate, which is readily decomposed into $H_2 + CO_2$ by non-methanogenic bacteria, can also be used as primary substrate for methane production (see Equation 4).

$$4HCOOH \rightarrow 3CO_2 + CH_4 + 2H_2O \qquad\qquad (4)$$

A likely alternative route for the hydrogen sink reaction may be acetogenesis from CO_2 by non-methanogenic bacteria, as indicated in Equation 5. This metabolic pathway occurs mainly in the hindgut whereas it is not significant in the rumen.

$$2CO_2 + 4H_2 \rightarrow CH_3COOH + 2H_2O \qquad\qquad (5)$$

Terminal electron acceptors other than CO_2 can also be utilised in an anaerobic ecosystem such as the rumen: the double bond of fumarate can be saturated by H_2 to produce succinate and then propionate; sulfate can be reduced to sulfite and sulfide; nitrate can be converted to nitrite; unsaturated fatty acids can also be saturated. However, these metabolic pathways, like Equation 5, are of only minor importance compared to methanogenesis (Equation 3), since together they represent less than 20% of the total use of metabolic hydrogen in the rumen. Methane is therefore a good indicator of fermentative activity (hydrogen production) while maintaining a low partial pressure of hydrogen in favour of microbial growth and microbial protein production (Demeyer and Van Nevel, 1975). Thus, any attempt to reduce methane production can have detrimental effects on the balance of ruminal microbes and their digestive and fermentative activities.

3. Impact of fermentation pattern on ruminal methanogenesis

Methane is produced in both the digestive compartments – rumen and hindgut – which house large microbial populations ($>10^{11}$ cells/g). The forestomach is the main site of methanogenesis since the hindgut accounts for about 10% of total methane production (Murray *et al.*, 1978; Hagemeister and Kaufmann, 1980), although ranging from as low as 4% (McGin *et al.*, 2006) to higher than 20% (Benchaar *et al.*, 1998). Thus, only factors controlling methane production in the rumen will be discussed here.

It is accepted that the amount of digested carbon in the rumen is balanced with all the end-products described in Equation 6.

Digested feed OM = volatile fatty acids (VFAs) + gases + microbial OM   (6)

More precisely, the following mean stoichiometric yield between the end products of rumen fermentation, set on a molar basis, has been proposed by Jouany *et al.* (1995) (see Equation 7).

50 glucose equivalents = 59 acetate (C2) + 23 propionate (C3) + 9 butyrate (C4) + 24 CH_4 + 53 CO_2 + 230 ATP   (7)

This indicates that methane production can vary within a broad range according to the pattern of fermentation. It is positively related to hydrogen-producing fermentative reactions and negatively related to hydrogen-using fermentative reactions, as shown below.

Hydrogen-producing reactions:

1 Glucose → 2 pyruvate + 4H (Embden-Meyerhof-Parnas pathway)
1 Pyruvate + H_2O → C2 + CO_2 + 2H

Hydrogen-using reactions:

$$Pyruvate + 4H \rightarrow C3 + H_2O$$
$$2C2 + 4H \rightarrow C4 + 2H_2O$$
$$CO_2 + 8H \rightarrow CH_4 + 2H_2O$$

Thus, the balance of hydrogen is positive when acetate or butyrate (+ 4H per mole of C2 and C4) are produced, and negative when propionate (-2H per mole of C3) is produced. As indicated previously, methanogenesis is the most efficient H-using pathway (-8H per mole of methane). Microbial cells can also fix some amount of metabolic hydrogen during synthesis since their oxidation-reduction status is negative (Demeyer and Van Nevel, 1975; Van Kessel and Russell, 1996). Figure 1 gives a global flow diagram of the fermentative pathways occurring in the rumen, including the exchanges of metabolic hydrogen.

Based on the assumption that on a molar basis, the amount of 2H produced (Hp) is equal to 2H used (Hu), Demeyer and Van Nevel (1975) proposed the following Equation (8) obtained from the previous reactions: 2 C2 + C3 + 4 C4 = 4 CH_4 + 2 C3 + 2 C4. Considering a mean recovery rate of metabolic hydrogen (estimated as the Hu/Hp ratio) of 0.90, then the previous equation makes it possible to estimate methane production from VFA production.

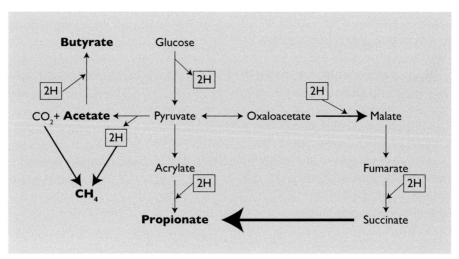

Figure 1. The major fermentative pathways in the rumen, including exchanges of metabolic hydrogen.

$$CH_4 = 0.45\ C2 - 0.275\ C3 + 0.40\ C4 \tag{8}$$

Equation 8 demonstrates that the amount of each VFA influences the production of methane. Acetate and butyrate promote methane production, whereas propionate formation can be considered as a true competitive pathway for hydrogen use in the rumen. Such theoretical calculations have been confirmed *in vitro*, where quantitative balances of end-products are reliably quantified. Methane production was accurately measured when molar proportions of individual VFAs were changed by adding monensin (Moss *et al.*, 2000). The results indicated that methane was not correlated to C2 alone ($r^2 = 0.029$) but was negatively correlated to C3 ($r^2 = 0.774$). The correlation between methane and C2/C3 ratio had approximately the same value ($r^2 = 0.772$). The addition of C4 to C2 in the previous ratio slightly improved the relationship ($r^2 = 0.778$), as shown in Figure 2. This result is consistent with the idea that propionate production and methanogenesis compete for regenerating oxidized co-factors in the rumen (Russell, 1998).

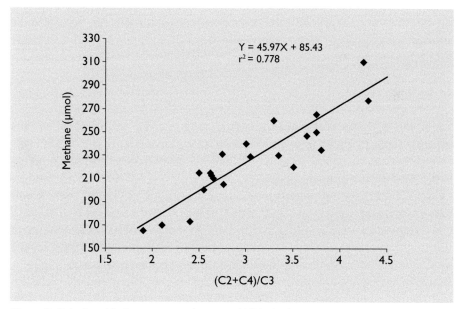

Figure 2. Relationship between methane and VFAs in the rumen.

4. Economic and environmental impact of enteric methane emissions by ruminants

All the methane generated in the rumen and in the hindgut is released into the atmosphere. It therefore impacts not only on a severe loss of energy for animals, but also, contributes to the global warming effect through its ability to absorb high-infrared radiations, especially those with long wavelengths.

4.1. Methane accounts for energy losses in ruminant production

Current measurements of methane production in respiratory chambers or with the SF_6-tracer technique (Johnson *et al.*, 1994) have been compiled into several ruminant databases worldwide. They have been used extensively to generate mathematical models estimating methane release from dietary and (or) animal factors. The international panel of scientists that published the IPCC report (2006) used the "conversion factor" (Y_m = % of feed gross energy (GE) converted to CH_4) to calculate the "emission factor" (EF = kg CH_4 head^{-1} year^{-1}) of each livestock category. The annual emission were then calculated by multiplying the EF for each category with the corresponding annual average population of animals in each category. *This conversion factor "Y_m" can be utilised to evaluate the energy lost by ruminants in the form of methane.*

4.1.1. Influence of diet

The mean Y_m values reported in Table 1 were derived from published data (Murray *et al.*, 1978; Johnson and Johnson, 1995; Lassey *et al.*, 1997; Judd *et al.*, 1999; Leuning *et al.*, 1999; Ulyatt *et al.*, 2002a,b, 2005; Lassey, 2007). The energy losses attributable to methane thus represent from 2% to 7% of GE, depending on animal category and dietary conditions. Losses are the lowest with cereal-rich or high-digestible grass diets, and the highest with high-fibre diets with low digestibility. This result is in good agreement with the stoichiometry of ruminal fermentation, since fibre digestion in the rumen stimulates acetate production whereas starch digestion increases propionate production. Figure 3, which shows typical energy flows in ruminants fed on different diets, indicates that methane contribution to DE can be as high as 14% of DE for extremely low-quality diets and as low as 3% in feedlot diets. Many models used to predict methane production have shown the same kind of influence of dietary components (see Table 2).

Table 1. IPCC conversion factors (Y_m).

Livestock category	Y_m (%)
Feedlot cattle receiving 90% or more concentrate	3.0 ± 1.0
Dairy cows (cattle and buffalo) and their young	6.5 ± 1.0
Other cattle and buffalo fed low-quality crop residues and by-products	6.5 ± 1.0
Other grazing cattle and buffalo	6.5 ± 1.0
Lambs (<1 year old)	4.5 ± 1.0
Mature sheep	6.5 ± 1.0

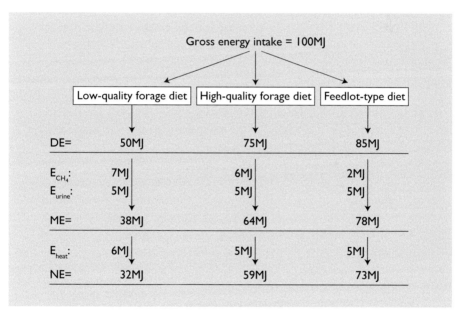

Figure 3. Typical energy flows and energy losses in ruminants fed different diets.

Sauvant and Giger-Reverdin (2007) applied the quantitative model called "meta-analysis" in statistics (St Pierre, 2001) to a database obtained from 87 experiments and 260 treatments to formulate regressions based on the influence of dietary proportion of concentrate ($0 < CO < 1$) on Y'_m ($Y'm = $ % of metabolisable energy converted to CH_4) (Figure 4). The best regression, $Y'_m = 13.2 + 4.2CO - 12.0CO^2$, was curvilinear with a maximum Y'_m value set at $CO = 0.175$ in the diet.

Table 2. A selection of models used to predict methane (M) emission by ruminants.

Authors	Animals	Models
Models based on diet composition		
Giger-Reverdin et al., 1992	Cattle	M(%DE) = 9.77 + 0.873CF [CF = crude fibre in % of OM]
Kirchgessner et al., 1995	Dairy cow	M(g/d) = 63 + 79CF(kg/d) + 10NFE + 26CP(kg/d) − 212Fat(kg/d) with a = 63 for cow and 16 for heifers [CF=crude fibre; NFE= N-free extracts; CP=crude protein]
Sauvant et al., 1999	Cattle	M(%GE) = 6.4 + 5.42C − 7.64C^2 rmse = 1.0 with C=% concentrate (0<C<1)
Giger-Reverdin et al., 2000	All ruminants	M(L/kgMSI) = 48.4 − 0.230ADL + 0.0116St − 0.155EE + 0.0224CP r = 0.75; etr = 7.2L/kgMS
		M(L/kgMS) = 55.9 − 0.235EE − 0.141ADL r = 0.52 ; etr = 3.4L/kgMS [St= starch % DM ; EE=ether extract % DM; CP=crude protein % DM]
Yan et al., 2000	Dairy + beef cattle	M(/GEI) = 0.0694T$_{ADFI}$/T$_{DMI}$ + 0.0522 r^2 = 0.463; rsd = 0.012 [T$_{ADFI}$ = total ADFI (kg/d); T$_{DMI}$ = total DMI (kg/d)]
		M(/GEI) = 0.0476S$_{ADFI}$/T$_{ADFI}$ + 0.0315 r^2 = 0.463; rsd = 0.011 [S$_{ADFI}$ = silage ADF intake (kg/d)]
Yates et al., 2000	Dairy cow	M(MJ/d) = 1.36 + 1.21VCDMI(kg) − 0.825DM(kg concentrates) + 12.8NDF/VCDMI(Kg) r^2 = 0.54, rmse = 1.97 [VCDMI= volatile corrected DMI]
Ellis et al., 2007	Beef cattle	M(MJ/d) = 2.94 + 0.0585MEI(MJ/d) + 1.44ADF(kg/d) − 4.16lignin(kg/d) rmspe = 14.4%
	Dairy cows	M(MJ/d) = 8.56 + 0.14forage(%) rmspe = 20.6%
Models including the level of feed intake		
Shibata et al., 1993	All ruminants	M(L/d) = -17.766+ 42.793DMI(kg/d) − 0.849DMI^2 r = 0.966
Yan et al., 2000	Dairy + beef cattle	M(MJ/d) = 0.0547GEI(MJ/d) + 3.2340 r^2 = 0.846 rsd = 3.016
		M(MJ/J) = 0.0714DEI(MJ/d) + 3.3180 r^2 = 0.841 rsd = 3.032
Giger-Reverdin et al., 2000	Growing cattle	M(L/kgDM) = 52.1 − 1.71DMI + 0.0267DMI^2 r = 0.52 etr = 7.6L/kg DM
Mills et al., 2003	Dairy cow	M(MJ/d) = 5.93 + 0.92DMI(kg/d) r^2 = 0.60; rmse = 1.82
Ellis et al., 2007	Dairy cows	M(MJ/d) = 3.23 + 0.81DMI(kg/d) rmspe = 25.6%
	Beef + dairy data	M(MJ/d) = 3.27 + 0.736DMI(kg/d) rmspe = 28.2%

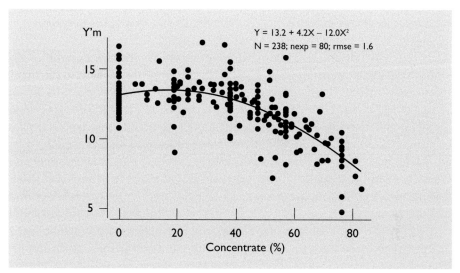

Figure 4. Influence of dietary proportion of concentrate on Y'_m *(Sauvant and Giger-Reverdin, 2007).*

Van Kessel and Russell (1969) showed that archaea found in the rumen are pH-sensitive, and that they lose the ability to use H_2 and thus have no energy supply for growth when the pH is below 5.5. They also indicated that the effects of both pH and propionate are independent and can be additive in cereal-rich diets that increase propionate proportion in the VFA mixture and lower the pH. The strict effect of pH alone leads to an accumulation of free H_2 in the gas phase, whereas propionate synthesis leads to a decrease in H_2 concentration.

Adding lipids to increase the energy content of diets for high-producing animals, especially for high-merit dairy cows, has a significant negative effect on the amount of methane produced per unit of dry matter intake (DMI), gross energy intake (GEI) or digestible energy intake (DEI) (McAllister *et al.*, 1996; Nelson *et al.*, 2001). This effect is explained by several factors: (1) a decrease in fibre digestion due a protective action of lipids surrounding the feed particles in the rumen, (2) an inhibiting effect of some fatty acids on archaea methanogens, and (3) the use, to some extent, of hydrogen to saturate certain unsaturated fatty acids (Demeyer and Doreau, 1999).

4.1.2. Influence of level of intake

As early as 1965, Blaxter and Clapperton showed that Y'$_m$ decreases significantly when the animals' intake levels increase. This effect has been confirmed in many recent experiments (Kirkpatrick *et al.*, 1997; Pelchen and Peters, 1998; Islam *et al.*, 2000; Yan *et al.*, 2002; Gabel *et al.*, 2003; Lovett *et al.*, 2005; McGinn *et al.*, 2006) and has been included in some models to predict methane emission by ruminants (see Table 2). The major effect of feeding level is explained by an accelerated turnover rate of feed particles out of the rumen (Owens and Goetsch, 1986). A reduction in methane production can be expected at lower feed residence times in the rumen, since ruminal digestion decreases and methanogenic bacteria are less able to compete with other H_2-using bacteria in these conditions. Furthermore, an increase in passage rate of feed particles in the rumen promotes propionate production and competitive pathways for hydrogen use (Isaacson *et al.*, 1975; Hoover *et al.*, 1984). According to Kennedy and Milligan (1978) and Okine *et al.*, (1989), there is a 30% decline in methane production when the ruminal passage of liquid and solid phases increases by about 60%. Mean retention time was shown to explain 28% of the variation in methane emissions (Okine *et al.*, 1989). As a consequence, an increase in feeding level induces lower methane losses calculated as a percentage of daily energy intake (Blaxter and Clapperton 1965; Blaxter and Wainman 1961; Moss *et al.*, 1995). Johnson and Johnson (1995) noted that methane losses expressed as the percentage of GEI decreased by 1.6 percentage units for each multiple of intake. Sauvant and Giger-Reverdin (2007) applied statistical meta-analysis to data from 101 experiments and 290 treatments in order to evidence the effect of feeding level (kg DM intake/100 kg of liveweight) on Y'$_m$. They reported that the best regression was linear, with a mean slope of 1.96 (Figure 5).

4.1.3. Influence of age of animals

Ruminants have an allometric evolution of microbial digestion and therefore of methane production, which is higher than 1 after the weaning period. Such evolution is explained by an allometric increase of rumen and hindgut volumes with age. Furthermore, it is strongly suggested that archaea methanogens need a long time (several months at least) to reach optimal concentration and activity in the rumen but there has been no confirmation to date on this point. Thus, it is considered that the amount of methane originating from the rumen increases during the production life of ruminants. For example, losses in energy due to methane accounts for about 3% GE in a 100-150-day-old bull calf and increases to about 7% in the same animal at 300 days old (Jentsch *et al.*, 1976; Vermorel *et al.*, 1980).

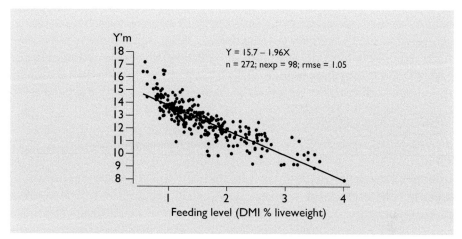

Figure 5. Influence of feeding level on Y'_m (Sauvant and Giger-Reverdin, 2007).

4.2. Impact of enteric methane on global warming

Accurate measurements of atmospheric gas composition have been carried out at the Scripps Institute of Oceanography (California), at the Mauna Lao volcano (Hawaii) and at many other places in the world. They have been completed by analyses of gases trapped in ice cores drilled mainly from the Antarctic and Artic and which extend back as far as 160,000 years. There is now clear evidence of changes in the composition of the greenhouse gases in the atmosphere, especially in the troposphere. While carbon dioxide attracts the most attention as a factor driving global warming, other gases such as methane, nitrous oxide and chlorofluorocarbons have also to be considered, even if their concentrations in the atmosphere remain low because of their high global warming power (GWP), which stems from their lifetime in the atmosphere and their capacity to absorb radiations. Thus, methane has a GWP which is 23 times higher than that of CO_2 (on weight basis) whereas its concentration is actually 210 times less. Methane accounts for 20% of the total radiative forcing by all the long-lived and globally mixed greenhouse gases, while carbon dioxide is estimated to represent 60% of the total radiative forcing effect.

Since 1750, global average atmospheric concentrations of methane have increased by 150%, from near 700 to 1,745 parts per billion by volume (ppbv) (IPCC, 2001). Over the past decade, although methane has continued to increase, the overall rate has slowed. The rising concentration of methane is closely correlated to the rising worldwide population, and currently about

60-70% of methane arises from anthropogenic sources with the remainder generated by natural sources (Duxbury *et al.*, 1993). Agriculture is considered to be responsible for more than 50% of anthropogenic emissions of methane (Table 3). Given that the greenhouse effect is accelerating, policymakers strongly recommend reducing the emissions of greenhouse gases, with methane being targeted due to its high GWP.

4.3. Mitigation scenarios for methane emissions from ruminants

Attempts have been made to reduce methane emissions by ruminants in order to both improve the energetic status of animals and control the current acceleration of the global warming effect. There are several different candidate strategies for methane mitigation. One consists in using dietary conditions or additives to control the microbial ecosystem in the rumen and stimulate

Table 3. Yearly major sources and sinks of methane at worldwide level.

Natural sources	
Natural wetlands, ocean and lakes	255 Tg
Termites	10 Tg
Total from natural sources	265 Tg
Anthropogenic sources	
Energy (oil, gas, coal)	130 Tg
Wetland rice fields	110 Tg
Enteric fermentation	80 Tg
Biomass burning	40 Tg
Landfills	40 Tg
Animal waste	25 Tg
Total from anthropogenic sources	425 Tg
Total emissions	690 Tg
Sinks of methane	
Soils	-40 Tg
Troposphere (OH radical oxidation)	-510 Tg
Stratosphere reactions	-50 Tg
Total sink	-600 Tg
Estimated CH_4 imbalance	+ 90 Tg

alternative pathways for hydrogen use; others are related to genetic selection of animals or animal production management strategies.

4.3.1. Control of microbial population and fermentation pattern in the rumen

Several halogenated methane analogues and related compounds (chloroform, chloral hydrate, amichloral, trichloroacetamide, trichloroethyladipate) have been tested *in vitro* as potent methane inhibitors (Van Nevel and Demeyer, 1995; Bauchop, 1967; Prins, 1965; Quaghebeur and Oyaert, 1971). However, some of them have led to liver damage and sometimes even death of animals when tested *in vivo* for several days (Lanigan *et al.*, 1978), and their effects were only maintained for short periods (Clapperton, 1974; 1977; Trei *et al.*, 1972). Bromochloromethane, which is known for its high antimethanogenic activity, also had only a transient effect (Sawyer *et al.*, 1974) but it has been suggested that co-infusing with a special delivery system (May *et al.*, 1995) or combining it with α-cyclodextrin (McCrabb *et al.*, 1997) would improve its stability and its activity for a prolonged period. 2-bromoethanesulfonic acid (BES), which is a bromine analogue of coenzyme F involved in methyl group transfer during the methanogenesis pathway, has proven efficient against archaea methanogens and methanogenesis (Martin and Macey, 1985; Wolfe, 1982) but yet again its effect is only transient (Van Nevel and Demeyer, 1995). 9,10-anthraquinone inhibits methanogenesis *in vitro* (Garcia-Lopez *et al.*, 1996) and suppresses methane production in lambs over a 19 day-period (Kung *et al.*, 1998). However, practical use of such chemicals seems unlikely because of their potential toxicity to animals and their possible transfer to edible animal products such as milk and meat. None of these additives are authorised for use in Europe. More recently, Wolin and Miller (2006) showed that statins currently used to treat hypercholesterolemia in humans through their inhibiting effect on HMG-CoA reductase could also be used to inhibit synthesis of mevalonate, which is a key precursor for isoprenoid synthesis by methanogens, since isoprenoid alcohols are essential compounds found in the unique membrane of *Archae*. The authors confirmed that statins have a specific inhibiting effect on rumen methanogens and did not affect the other important bacterial species that ferment cellulose, starch and other plant polysaccharides (Miller and Wolin, 2001). However, statins are far too expensive to be considered for future use as additives in animal production.

Many studies carried out on ionophore antibiotics such as monensin have shown that these additives can depress methane production *in vitro* and *in vivo* via a direct effect on archaea methanogens and by shifting the fermentation pattern toward an increase in propionate at the expense of acetate (Jouany and Senaud, 1978; Chen and Wolin, 1979; Caffarel-Mendez *et al.*, 1986). Van Nevel and Demeyer (1995) found that monensin led to a mean decrease in methane production of 25% when averaged over 6 *in vivo* studies. However, the use of any antibiotics as feed additives in Europe has been forbidden after 1st January 2006 (EC Regulation No. 1831/2003 of the European Parliament and the Council of 22 September 2003), making these compounds ineligible for use as antimethanogens.

4.3.2. Effect of yeast addition

Only few experiments have been carried out to assess the effect of live yeast addition to the diet of ruminants on methane emissions. Early results obtained by Mutsvangwa *et al.* (1992) indicated that Yea-Sacc[1026] decreased *in vitro* methane production. More recent research reported that another strain of live yeast *Saccharomyces cerevisiae* (PMX70SBK Sa Agri) decreased methane significantly during *in vitro* digestion of alfalfa hay (Lynch and Martin, 2002). Such effect may be due to the mobilisation of metabolic hydrogen by acetogenic bacteria to produce acetate as suggested by Chaucheyras *et al.* (1995). However, there is limited information to demonstrate the effect of these additives *in vivo*. For example, a study conducted on 16 holstein steers in a latin square design showed no significant effect of *S. cerevisiae* (Levucell SC[1077]) on methane emissions (McGinn *et al.*, 2004). A new strategy with associated twin strains of *S. cerevisiae* (Yea-Sacc strains 8417 and 1026) has been suggested by Lila *et al.* (2004), which showed a significant reduction of methane *in vitro* but the results have not yet been validated *in vivo*.

4.3.3. Propionate enhancers

Callaway and Martin (1996), Martin (1998) and Lopez *et al.* (1999) suggested that dicarboxylic acids such as malate or fumarate may alter rumen fermentations in a similar way to ionophores but without the disadvantages of antibiotics. These compounds act as propionate precursors in the rumen: they enter metabolic pathway through hydrogen fixation, eventually leading to a decrease in methane production. Based on a hydrogen fixation efficacy by malate or fumarate of roughly 60%, Newbold and Rode (2006) calculated that it would be necessary to supplement diets with more than 2 kg of pure acids to get a 10% decrease in methane production. This kind of amount of pure acids would have a detrimental impact on rumen pH and would be too onerous to be realistically useable in ruminant feed strategies. Recently, Wallace *et al.* (2006) observed that encapsulated fumaric acid given to growing lambs at a rate of 10% of intake led to a 75% decrease in methane production whilst improving feed conversion ratio by 20%. Callaway *et al.* (1997) suggested supplying malate naturally through selected alfalfa cultivars, and the authors indicated that the desired efficient dose of malate could be obtained simply by incorporating 6.0 kg of alfalfa at 42 days of maturity into dairy cow diets. Although dicarboxylic acids have not yet been allowed as feed additives, they are considered as naturally-occurring substances since they are major metabolic intermediates of the citric acid cycle and are therefore commonly found in plant and animal tissues.

Over the last decade, many experiments have been carried out on the potential use of essential oils to reduce methanogenesis (see Kamra *et al.*, 2006; Jouany and Morgavi, 2007). However, no essential oils have yet obtained authorisation in the EU and the persistency of effect over a long period has still not been demonstrated.

4.3.4. Addition of lipids; specific effect of some fatty acids

A commonly-deployed strategy for increasing the energy density of diets given to high-merit animals is to add lipids. In contrast to dicarboxylic acids, they do not alter pH in the rumen and can both decrease methane production (Jouany, 1994) and improve the nutritive quality of animal product if the lipids administered are rich in unsaturated fatty acids (Chilliard *et al.*, 2001). However, they can also reduce fibre digestion when added at a rate higher than 6% of the diet (Ikwegbu and Sutton, 1982).

Their efficacy on methanogenesis depends on the type and the amount of lipids added. Polyunsaturated fatty acid-rich lipids are more effective than saturated or monounsaturated fatty acid-rich lipids. From a practical standpoint, linseeds, which are rich in 9,12,15-18:3 (linolenic acid), have been shown to decrease methane by 10% on growing lambs when added at a rate of 2.5% and by 38% in adult sheep at maintenance fed on grass hay supplemented with 5% linseeds (Czerkawski *et al.*, 1966). The amount of hydrogen used to saturate these fatty acids means it has no significant influence on the methane decrease. Unsaturated fatty acids would have a direct negative effect on rumen microbes both by limiting the intensity of fermentation (hydrogen production) and by inhibiting archaea methanogens (hydrogen use). Saturated medium-chain fatty acids such as lauric and myristic acids present in copra or palm oil are also have anti-methanogenic properties (Soliva *et al.*, 2004; Machmüller *et al.*, 2003). Contrary to most other additives, lipids exert long-term anti methane actions.

4.3.5. Genetic selection of animals

The current genetic selection based on high feed conversion efficiency will logically result in animals presenting more efficient in energy use and thus producing less methane (Arthur *et al.*, 2001). Despite this selection drive, individual within-herd differences in methane emission can be as high as 30% to 60% (Demeyer 1991; Lassey *et al.*, 1997, Lassey, 2007). This is due to physiological differences inducing different feeding characteristics (level and rate of intake) and between-individual differences in rumen microbial populations (protozoa, bacteria and fungi), even when fed the same diet. Pinares-Patino *et al.* (2003) and Goopy *et al.* (2006) identified low methane-emitters and high methane-emitters but the contradictory results obtained between the authors made it impossible to conclude on whether animal status was transient or permanent. More research must be undertaken in the future to examine the real possibility of a methane-based genetic selection trait.

4.3.6. Influence of the animal production system

There is no doubt that increasing animal productivity reduces the amount of enteric methane released per unit of animal product (milk or meat). Sauvant (1993) compared two groups of cows producing the same total quantity of milk – either 60 cows producing 4,000 kg milk each or 24 cows producing 10,000 kg milk each. The high-merit cows produced more absolute values of methane but due to their much higher milk yield, the relative values of methane per litre of milk were significantly lower. Thus, the amounts of methane produced per litre of milk were 28 g and 15 g in the extensive system and the intensive system of production respectively.

Pinares-Patino *et al.*, (2006) indicated that the stocking rate of animals on pasture (1.2 vs. 2.1) did not change individual methane emissions (total emissions or emissions per unit of liveweight) by heifers measured during late spring, mid summer, late summer and early autumn for the 2002 and 2003 grazing seasons. Both absolute methane emission (gd^{-1}) and Y'_m were consistently related to GEI and DEI. The degrees of the relationships, however, differed between systems and between grazing seasons.

For beef cattle, the amount of methane emitted per unit of carcass weight increases with animal age. Thus, fattening a bull until 19 months (700 kg) induces 580 l of methane per kg of carcass, whereas fattening a steer until 40 months

(690 kg) generates 1040 l (Vermorel, 1995). Here again, the intensification of animal production should foster lower methane production.

Regarding dairy cows, it is recommended to increase the number of lactations of each animal and decrease the culling rate of heifers in order to reduce the huge impact of the non-producing period, i.e., before the 1st lactation, on methane production. Johnson *et al.* (2002) calculated that a 10% decrease in culling leads to a 5% decrease in methane production. However, slowing down the culling rate will decrease the mean level of milk production in the herd and result in higher methane production in terms of per unit of milk.

5. Conclusion

Ruminant methanogenesis is a natural process driven by rumen microbes during feed digestion. It is generally used in *in vitro* systems as an indicator of optimised microbial functions. While it is impossible to believe that methane production can be totally eliminated, a 30% reduction in methane emissions is a distinct possibility. Beyond this point, some basic ruminal functions such as cellulolysis will be significantly impaired. In terms of antimethanogenic feed additives, there are only a few that fit with the three main requirements, i.e., efficient over long periods, non-toxic for animals, the environment and consumers, and cheap enough for standard use in animal feeds. The intensification of animal production is another possibility for reducing methane production per unit of product. Animal production systems based on shortening the production life of fattening cattle while at the same time increasing the number of lactations of dairy cows could also be used to reduce methane emissions per unit of animal product.

However, this paper only covers enteric gases and the solutions put forward must be evaluated in more integrated systems. For example, more intensive animal production systems will require more cereals to be included in the diet, which in turn will induce greenhouse gases generated through crop production. In contrast, grazing will have a positive effect on the amount of CO_2 fixed on pasture (Soussana *et al.,* 2007). Also, more greenhouse gases are produced from liquid manure stored for long periods under intensive indoor animal production systems than under grazing systems and comparison between the two systems indicates that the indoor system produces more total methane than the grazing system (Johnson *et al.,* 2000).

References

Arthur, P.F., J.A. Archer, D.J. Johnston, R.M. Herd, E.C. Richardson, and P.F. Parnell, 2001. Genetic and phenotypic variance and covariance components for feed intake, feed efficiency, and other post-weaning traits in Angus cattle. *Journal of Animal Science* **79:**2805-2811.

Bauchop, T., 1967. Inhibition of rumen methanogenesis by methane analogues. *Journal of Bacteriology* **94:**171-175.

Benchaar, C., J. Rivest, C. Pomar and J. Chiquette, 1998. Prediction of methane production from dairy cows using existing mechanistic models and regression equations. *Journal of Animal Science* 76:617-627.

Blaxter, K.L. and J.L. Clapperton, 1965. Prediction of the amount of methane produced by ruminants. *British Journal of Nutrition* **19:**511-522.

Blaxter, K.L. and F.W. Wainman, 1961. The utilization of food by sheep and cattle. *Journal of Agricultural Science* 57:419-425.

Caffarel-Mendez, S., J.-P. Jouany and C. Demuynck, 1986. Etude in vitro de quelques antibiotiques ionophores et de certains dérivés. I- Action sur les produits de la fermentation dans le rumen. *Reproduction Nutrition Development* **26:**1295-1303.

Callaway, T.R. and S.A. Martin, 1996. Effect of organic acid and monensin treatment on in vitro mixed ruminal microorganism fermentation of cracked corn. *Journal of Animal Science* **74:**1982-1989.

Callaway, T.R., S.A. Martin, J.L. Wampler, N.S. Hill and G.M. Gill, 1997. Malate content of forage varieties commonly fed to cattle. *Journal Dairy Science* **80:**1651-1655.

Chaucheyras, F., G. Fonty, G. Bertin and Gouet, P. 1995. In vitro hydrogen utilization by a ruminal acetogenic bacterium cultivated alone or in association with an archaea methanogen is stimulated by a probiotic strain of Saccharomyces cerevisiae. *Applied and Environmental Microbiology* **61:**3466-3467.

Chen, M., and M.J. Wolin, 1979. Effect of monensin and lasalocid sodium on the growth of methanogenic and rumen saccharolytic bacteria. *Applied and Environmental Microbiology* **38:**72-77.

Chilliard, Y., A. Ferlay and M. Doreau, 2001. Effect of different types of forages, animal fats or marine oils in cow's diet on milk fat secretion and composition, especially conjugated linoleic acid (CLA) and polyunsaturated fatty acids. *Livestock Production Science* **70:**31-48.

Clapperton, J.L., 1974. The effect of trichloroacetamide chloroform and linseed oil given into the rumen of sheep on some of the end products of rumen digestion. *British Journal of Nutrition* **32:**155-161.

Clapperton, J.L., 1977. The effect of methane-suppressing compound trichloroethyl adipate on rumen fermentation and the growth of sheep. *Animal Production* **24:**169-181.

Czerkawski, J.W., K.L. Blaxter and F.W. Wainman, 1986. The metabolism of oleic, linoleic and linolenic acids by sheep with reference to their effects on methane production. *British Journal of Nutrition* **20**:349-362.

Demeyer, D.I., 1991. Quantitative aspects of microbial metabolism in the rumen and the hindgut. In: *Rumen Microbial Metabolism and Ruminant Digestion* (Ed. J.-P. Jouany). INRA Editions, Versailles, France. pp. 217-237.

Demeyer, D.I. and C.J. Van Nevel, 1975. Methanogenesis, an integrated part of carbohydrate fermentation, and its control. In: *Digestion and Metabolism in the Ruminant* (Eds. I.W McDonald and A.C.I. Warner). The University of New England Publishing Unit, Armidale, Australia. pp. 366-382.

Demeyer, D.I. and M. Doreau, 1999. Targets and procedures for altering ruminant meat and milk lipids. *Proceedings of the Nutrition Society* 58: 593-607.

Duxbury, J.M., L.A. Harper and A.R. Mosier, 1993. Contributions of agroecosystems to global climate change. Proceedings of the Symposium on Agricultural Ecosystem Effects on Trace Gases and Global Climate Change, Denver, Colorado, USA. pp1-18.

Ellis, J.L., E. Kebreab, N.E. Odongo, B.W. McBride, E.K. Okine and J. France, 2007. Prediction of Methane Production from Dairy and Beef Cattle. *Journal of Dairy Science* **90**:3456-3466.

Gabel, M., B. Pieper, K. Friedel, M. Radke, A. Hageman, J. Voigt and S. Kuhla, 2003. Influence of nutrition level on digestibility in high yielding cows and effects on energy evaluation systems. *Journal of Dairy Science* **86**:3992-3998.

Garcia-Lopez, P.M., L. Kung, and J.M.L. Odhom, 1996. In vitro inhibition of microbial methane production by 9,10-anthraquinone. *Journal of Animal Science* **74**:2276-2284.

Giger-Reverdin, S., M. Vermorel and D. Sauvant, 1992. Facteurs de variation de la production de méthane au cours de la digestion des aliments composés chez les ruminants. *Annales de Zootechnie* **41**:37-38.

Giger-Reverdin, S., D. Sauvant, M. Vermorel and J.-P. Jouany, 2000. Empirical modeling of methane losses from ruminants. *Rencontres Recherches Ruminants* **7**:187-190.

Goopy, J.P., R.S. Hegarty and R.C. Dobos, 2006. The persistence over time of divergent methane production in lot fed cattle. *International Congress Series* **1293**:111-114.

Hagemeister, H. and W. Kaufman, 1980. Nährstoff-Fermentation im Dickdarm der Wiederkäuer und Konsequenzen für die Messung der Proteinverdaulichkeit. *Übersichten Zur Tierernährung* **8**:101-122.

Hoover, W.H., C.R. Kincaid, G.A. Varga, W.V. Thayne and L.L. Junkins, 1984. Effects of solid and liquid flows on fermentation in continuous cultures. IV. pH and dilution rate. *Journal of Animal Science* **58**:692-699.

Ikwegbu, O.A. and J.D. Sutton, 1982. The effect of varying the amount of linseed oil supplementation on rumen metabolism in sheep. *British Journal of Nutrition* **48:**365-375.

IPCC Report, 2001. Climate Change: The Scientific basis. Contribution of Working Group to the 3rd Assessment Report of the Intergovernmental Panel on Climate Change [www.ipcc.ch/pub/online-htm].

IPCC Report, 2006. Emissions from livestock and manure management. In: *IPCC Guidelines for National Greenhouse Gas Inventories. Volume 4. Agriculture, Forestry and Other Land Use* (Eds. S. Eggleston, L. Buendia, K. Miwa, T. Ngara and K. Tanabe). Published by the Institute for Global Environmental Strategies, Hayama, Japan, pp. 1-87.

Isaacson, H.R., F.C. Hinds, M.P. Bryant and F.C. Owens, 1975. Efficiency of energy utilization by mixed rumen bacteria in continuous culture. *J. Dairy Science* **58:**1645-1659.

Islam, M., H. Abe, Y. Hayashi and F. Terada, 2000. Effects of feeding Italian ryegrass with corn on rumen environment, nutrient digestibility, methane emission, and energy and nitrogen utilization at two levels by goats. *Small Ruminant Research* **38:**165-174.

Jentsch, W., H. Wittenburg and R. Schiemann, 1976. Utilization of feed energy by growing bulls. Comparative studies of the digestibility of 44 rations and physiological indices of the rumen in young bulls and adult sheep. *Archiv für Tierernährung* **26:**575-585.

Johnson, K.A. and D.E. Johnson, 1995. Methane emissions from cattle. *Journal of Animal Science* **73:**2483-2492.

Johnson, D.E., H.W. Phetteplace and M.J. Ulyatt, 2000. Variation in the proportion of methane of total greenhouse gas emissions from US and NZ dairy production systems. Proceedings of the 2nd International Symposium on Methane Mitigation, Novosibirsk, Russia, pp. 249-254.

Johnson, D.E., H.W. Phetteplace and A.F. Seild, 2002. Methane, nitrous oxide and carbon dioxide emissions from ruminant livestock production systems. In: Greenhouse Gases and Animal Agriculture (Eds. J. Takahashi and B.A.Young). Elsevier Science BV, Amsterdam, The Netherlands, pp. 77-85.

Johnson, K., M. Huyler, H. Westberg, B. Lamb and P. Zimmermann, 1994. Measurement of methane emissions from ruminant livestock using a SF6 tracer technique. *Environmental Science and Technology* **28:**359-362.

Jouany, J.-P., 1994. Manipulation of microbial activity in the rumen. *Archives of Animal Nutrition (Berlin)* **46:**133-153.

Jouany, J.-P. and D.P. Morgavi, 2007. Use of "natural" products as alternatives to antibiotic feed additives in ruminant production. *Animal* (in press).

Jouany, J.-P. and J. Senaud, 1978. Utilisation du monensin dans la ration des ruminants. II- Effets sur les fermentations et la population microbienne du rumen. *Annales de Zootechnie* **27**:61-74.

Jouany, J-P., L. Broudiscou, R.A. Prins and S. Komizarzuk-Bony, 1995. Métabolisme et nutrition de la population microbienne du rumen. *Nutrition des Ruminants Domestiques; Ingestion et Digestion.* (Eds. R. Jarrige, Y. Ruckebusch, C. Demarquilly, M.H. Farce and M. Journet). INRA Editions, Versailles, France, pp. 349-381.

Judd, M.J., F.M. Kelliher, M.J. Ulyatt, K.R. Lassey, K.R. Tate, I.D. Shelton, M.J, Harvey and C.F. Walker, 1999. Net methane emissions from grazing sheep. *Global Change Biology* **5**:647-657.

Kamra, D.N., N. Agarwal and L.C. Chaudhary, 2006. Inhibition of ruminal methanogenesis by tropical plants containing secondary compounds. *International Congress Series* **1293**:156-163.

Kennedy, P.M. and L.P. Milligan, 1978. Effect of cold exposure on digestion, microbial synthesis and nitrogen transformation in sheep. *British Journal of Nutrition* **39**:105-117.

Kirchgessner, M., W. Windisch and H.L. Müller, 1995. Nutritional factors for the quantification of methane production. In: *Ruminant Physiology: Digestion, Metabolism, Growth and Reproduction* (Eds. W.v. Engelhardt, S. Leonhard-Marek, G. Breves and D. Giesecke). Ferdinand Enke Verlag, Stuttgart, Germany, pp. 333-348.

Kirkpatrick, D.E., R.W.J. Steen and E.F. Unsworth, 1997. The effect of differing forage: concentrate ratio and restricting feed intake on the energy and nitrogen utilization by beef cattle. *Livestock Production Science* **51**:151-164.

Kung, L., A.O. Hession, and J.P. Bracht, 1998. Inhibition of sulfate reduction to sulfite by 9,10-anthraquinone in in vitro ruminal fermentations. *Journal of Dairy Science* **81**:2251-2256.

Lanigan, G.W., A.L. Payne and J.E. Peterson, 1978. Antimethanogenic drugs and *Heliotropium europaeum* poisoning in penned sheep. *Australian Journal of Agricultural Research* **29**:1281-1291.

Lassey, K.R., M.J. Ulyatt, R.J. Martin, C.F. Walker and I.D. Shelton, 1997. Methane emissions measured directly from grazing livestock in New Zealand. *Atmospheric Environment* **31**:2905-2914.

Lassey, K.R., 2007. Livestock methane emission: From the individual grazing animal through national inventories to the global methane cycle. *Agricultural and Forest Meteorology* **102**:120-132.

Leuning, R., S.K. Baker, I.M. Jamie, C.H. Hsu, L. Klein, O.T. Denmead and D.W.T. Griffith, 1999. Methane emission from free-ranging sheep: a comparison of two measurement methods. *Atmospheric Environment* **33**:1357-1365.

Lila, Z.A., N. Mohammed, T. Yasui, Y. Kurokawa, S. Kanda and H. Itabashi, 2004. Effects of a twin strain of *Saccharomyces cerevisiae* live cells on mixed ruminal microorganism fermentation in vitro. *Journal of Animal Science* **82**:1847-1854.

Lopez, S., C., Valdez, C.J., Newbold, and R.J. Wallace, 1999. Influence of sodium fumarate addition on rumen fermentation in vitro. *British Journal of Nutrition* **81**:59-64.

Lovett, D.K., L.J. Stack, S. Lovell, J. Callan, B. Flynn, M. Hawkins and F.P. O'Mara, 2005. Manipulating enteric methane emissions and animal performance of late-lactation dairy cows through concentrate supplementation at pasture. *Journal of Dairy Science* **88**:2836-2842.

Machmüller, A., C.R. Soliva and M. Kreuer, 2003. Methane-suppressing effect of myristic acid in sheep as affected by dietary calcium and forage proportion. *British Journal of Nutrition* **90**:529-540.

Martin, S.C., 1998. Manipulation of ruminal fermentation with organic acids: a review. *Journal of Animal Science* **76**:3123-3132.

Martin, S.A. and J.M. Macey, 1985. Effects of monensin, pyromellitic diimide and 2-bromoethanesulfonic acid on rumen fermentation in vitro. *Journal of Animal Science* **68**:2142-2149.

May, C., A.I. Payne, P.L. Stewart and J.A. Edgar, 1995. A delivery system for agents. International Patent Application. PCT/AU95/00733.

McAllister, T.A., E.K. Okine, G.W. Mathison and K.J. Cheng, 1996. Dietary, environmental and microbiological aspects of methane production in ruminants. *Canadian Journal of Animal Science* **76**:231-243.

McCrabb, G.J., K.T. Berger, T. Magner, C. May and R.A. Hunter, 1997. Inhibiting methane production in Brahman cattle by dietary supplementation with a novel compound and the effects on growth. *Australian Journal of Agricultural Research* **48**:323-329.

McGinn, S.M., K.A. Beauchemin, T. Coates and D. Colombatto, 2004. Methane emissions from beef cattle: effect of monensin, sunflower oil, enzymes, yeast, and fumaric acid. *Journal of Animal Science* **82**:3346-3356.

McGinn, S.M., K.A. Beauchemin, A.D. Iwaasa and T.A. McAllister, 2006. Assessment of the sulfur hexafluoride (SF6) tracer technique for measuring enteric methane emissions from cattle. *Journal of Environmental Quality* **35**:1686-1691.

Miller, T.L., 1995. Ecology of methane production and hydrogen sinks in the rumen. In: *Ruminant Physiology: Digestion, Metabolism, Growth and Reproduction* (Eds. W.v. Engelhardt, S. Leonhard-Marek, G. Breves and D. Giesecke). Ferdinand Enke Verlag, Stuttgart, Germany, pp.317-331.

Miller, T.L. and M.J. Wolin, 2001. Inhibition of growth of methane-producing bacteria of the ruminant forestomach by hydroxymethylglutaryl-SCoA reductase inhibitors. *Journal of Dairy Science* **84**:1445-1448.

Mills, J.A.N., E. Kebreab, C.M. Yates, L.A. Crompton, S.B. Cammell, M.S. Dhanoa, R.E. Agnew and J. France, 2003. Alternative approaches to predicting methane emissions from dairy cows. *Journal of Animal Science* **81**:3141-3150.

Moss, A.R., D.I. Givens and P.C. Garnsworthy, 1995. The effect of supplementing grass silage with barley on digestibility, in sacco degradability, rumen fermentation and methane production in sheep at two levels of intake. *Animal Feed Science and Technology* **55**:9-33.

Moss, A.R., J.-P. Jouany and J. Newbold, 2000. Methane production by ruminants: its contribution to global warming. *Annales de Zootechnie* **49**:231-253.

Murray, R.M., A.M. Bryant and R.A. Leng, 1978. Methane production in the rumen and lower gut of sheep given lucerne chaff: effect of level of intake. *British Journal of Nutrition* **39**:337-345.

Mutsvangwa, T., I.E. Edwards, J.H. Topps and G.F.M. Paterson, 1992. The effect of dietary inclusion of yeast culture (Yea-Sacc) on patterns of rumen fermentation, food intake and growth of intensively fed bulls. *Animal Production* **55**:35-41.

Nelson, M.L., H.H. Westberg and S.M. Parish, 2001. Effects of tallow on the energy metabolism of wethers fed barley finishing diets. *Journal of Animal Science* **79**:1892-1904.

Newbold, C.J. and L.M. Rode, 2006. Dietary additives to control methanogenesis in the rumen. *International Congress Series* **1293**:138-147.

Okine, E.K., G.W. Mathison and R.T. Hardin, 1989. Effects of changes in frequency of reticular contractions on fluid and particulate rates in cattle. *Journal of Animal Science* **67**:3388-3396.

Owens, F.N., and A.L. Goetsch, 1986. Digesta passage and microbial protein synthesis. In: *Control of Digestion and Metabolism in Ruminants*. (Eds. L.P. Milligan, Grovum, W.L. and Dobson, A.). A Reston Book Prentice Hall, Englewood Cliffs, NJ, USA, pp. 196-223.

Pelchen, A., and K.J. Peters, 1998. Methane emissions from sheep. *Small Ruminant Research* **27**:137-150.

Pinares-Patiño, C., P. D'Hour and C. Martin, 2003. Methane emissions by Charolais cows grazing a monospecific pasture of timothy at four stages of maturity. *Canadian Journal of Animal Science* **83**:769-777.

Pinares-Patiño, C., P. D'Hour, J.-P. Jouany and C. Martin, 2006. Effects of stocking rate on methane and carbon dioxide emissions from grazing cattle. *Agriculture, Ecosystems and Environment* **121**:30-46.

Prins, 1965. Action of chloral hydrate on rumen microorganisms in vitro. *Journal of Dairy Science* **48**:991-993.

Quaghebeur, D. and W. Oyaert, 1971. Effect of chloral hydrate and related compounds on the activity of several enzymes in extracts of rumen microorganisms. *Zentralblatt fur Veterinarmedizin* **18**:417-427.

Russell, J.B., 1998. The importance of pH in the regulation of ruminal acetate to propionate ratio and methane production in vitro. Journal of Dairy Science **81:**3222-3230.

Sauvant, D., 1993. La production de methane dans la biosphere: le rôle des animaux d'élevage. *Courrier de la Cellule Environnement INRA* **18:**67-70.

Sauvant, D. and S. Giger-Reverdin, 2007. Empirical modelling by meta-analysis of digestive interactions and CH$_4$ production in ruminants. In: *Energy and Protein Metabolism and Nutrition* (Ed. I. Ortigues-Marty). Wageningen Academic Publishers, Wageningen, The Netherlands, pp561-562.

Sauvant, D., J-P. Jouany, S. Giger-Reverdin, M. Vermorel and G. Fonty, 1999. Methane production by ruminants: analysis of process, quantitative modelling, balance, spacialisation and possibility of reducing the production. *Comptes Rendus de l'Académie d'Agriculture de France* **85:** 70-86.

Sawyer, M.S., W.H. Hoover and C.J. Sniffen, 1974. Effects of a ruminal methane inhibitor on growth and energy metabolism in the ovine. *Journal of Animal Science* **38:**904-914.

Shibata, M., F. Terada, M. Kurihara, K. Nishida and K. Iwasaki, 1993. Estimation of methane production. *Animal Science and Technology* **64:**790-796.

Soliva, C.R., L. Meile, A. Cielak, M. Kreuzer and A. Machmüller, 2004. Rumen simulation technique study on the interactions of dietary lauric and myristic acid supplementation in suppressing ruminal methanogenesis. *British Journal of Nutrition* **92:**689-700.

Soussana, J.F., V. Allard, K. Pilegaard, P. Ambus, C. Amman, C. Campbell, E. Ceschia, J. Clifton-Brown, S. Czobel, R. Domingues, C. Flechard, J. Fuhrer, A. Hensen, L. Horvath, M. Jones, G. Kasper, C. Martin, Z. Nagy, A. Neftel, A. Raschi, S. Baronti, R.M. Rees, U. Skiba, P. Stefani, G. Manca and M. Sutton, 2007. Full accounting of the greenhouse gas (CO$_2$, N$_2$O, CH$_4$) budget of nine European grassland sites. *Agriculture, Ecosystems and Environment* **121:**121-134.

St Pierre, N.R., 2001. Integrating quantitative findings from multiple studies using mixed model methodology. *Journal of Dairy Science* **84:**741-755.

Trei, J.E., G.C. Scott and R.C. Parish, 1972. Influence of methane inhibition on energetic efficiency of lambs. *Journal of Animal Science* **34:**510-515.

Ulyatt, M.J., K.R. Lassey, I.D. Shelton and C.F. Walker, 2002a. Seasonal variation in methane emission from dairy cows and breeding ewes grazing ryegrass/white clover pasture in New Zealand. *New Zealand Journal of Agricultural Research* **45:**217-226.

Ulyatt, M.J., K.R. Lassey, I.D. Shelton and C.F. Walker, 2002b. Methane emission from dairy cows and wether sheep fed subtropical grass-dominant pastures in mid summer in New Zealand. *New Zealand Journal of Agricultural Research* **45:**227-234.

Ulyatt, M.J., K.R. Lassey, I.D. Shelton and C.F. Walker, 2005. Methane emission from sheep grazing four pastures in late summer in New Zealand. *New Zealand Journal of Agricultural Research* **48**:385-390.

Van Kessel, J.S. and J.B. Russell, 1969. The effect of pH on ruminal methanogenesis. *FEMS Microbiology Ecology* **20**:205-210.

Van Nevel, C.J. and D.I. Demeyer, 1995. Feed additives and other interventions for decreasing methane emissions. In: *Biotechnology in Animal Feeds and Animal Feeding* (Eds. R.J. Wallace and A. Chesson). VCH, Weinheim, Germany, pp 329-349.

Vermorel, M., 1995. Emissions annuelles de methane d'origine digestive par les bovines en France. Variations selon le type d'animal et le niveau de production. *INRA Production Animale* **8**:265-272.

Vermorel, M., J.C. Bouvier and Y. Geay, 1980. Utilization of feed energy by ruminant calves; effects of milk intake, feeding level and age. *Annales de Zootechnie* **29**:65-86.

Wallace, R.J., T.A. Wood, A. Rowe, J. Price, D.R. Yanez, S.P. Williams and C.J. Newbold, 2006. Encapsulated fumaric acid as a mean of decreasing ruminal methane emissions. *International Congress Series* **1293**:148-151.

Wolfe, R.S., 1982. Biochemistry of methanogenesis. *Experimentia* 38:198-200.

Wolin, M.J. and T.L. Miller, 2006. Control of rumen methanogenesis by inhibiting the growth and activity of methanogens with hydroxymethylglutaryl-SCoA inhibitors. *International Congress Series* **1293**:131-137.

Yan, T., R.E. Agnew, F.J. Gordon and M.G. Porter, 2000. Prediction of methane energy output in dairy and beef cattle offered grass silage-based diets. *Livestock Production Science* **64**:253-263.

Yan, T., R.E. Agnew and F.J. Gordon, 2002. The combined effects of animal species (sheep vs cattle) and level of feeding on digestible and metabolizable energy concentrations in grass-based diets of cattle. *Animal Science* **75**:141-151.

Yates, C.M., S.B. Cammell, J. France and Beever, D.E. 2000. Prediction of methane emissions from dairy cows using multiple regression analysis. *Proceedings of the British Society of Animal Science,* pp. 94 (abstract).

Rate of nitrogen and energy release in the rumen and effects on feed utilisation and animal performance

L.A. Sinclair
Harper Adams University College, Edgmond, Newport, Shropshire TF10 8NB, United Kingdom; lsinclair@harper-adams.ac.uk

1. Introduction

Feed supply for dairy and beef cattle is currently going through a period of considerable instability and uncertainty. For example, in the UK rapeseed meal and soybean meal have risen from £108 and £148/t a year ago to current market values of around £160 and £216/t respectively (at the time of writing). Future increased demand for crops for human use and the uncertainty regarding the use of crops for fuel generation will result in a greater reliance being placed on efficient utilisation of forages and for reliable raw material nutrient characteristics to be used within ration formulation packages. As rumen fermentation can supply 70-100% of a ruminant animals protein supply and 70-85% of the energy supply can be absorbed as volatile fatty acids (VFA), the main end-product of microbial fermentation, optimising microbial growth is of obvious importance. Within this, maximising the utilisation of rumen degradable protein by converting it efficiently to microbial protein is a key objective in beef and dairy cow feeding strategies.

Ration formulation systems for ruminants generally have the common approach of considering energy and protein to the rumen and tissues on a 24-h basis. This approach is convenient in relation to human work-schedules but it is questionable whether rumen microbes or tissues operate to such a short time-scale. Additionally, such a time-scale does not consider periods of excess nutrients and the consequences of this on whole body metabolism, health and fertility. This review will consider the effects of rumen degradable protein supply to the rumen and how microbial protein production can be optimised to enhance animal performance, feed utilisation and reduce feed costs.

2. Interaction between rate of energy and nitrogen supply to the rumen on microbial metabolism.

Despite feed formulation systems considering nutrient supply to the rumen and tissues on a daily basis, there is a body of data to support that considering nutrient supply on a shorter time-scale can affect animal performance. For example, studies in pigs have demonstrated that an unmatched or asynchronous supply of glucose and amino acids within a day increases amino acid oxidation, resulting in a substantial reduction in protein utilisation (Van den Borne *et al.*, 2007). Similarly, in pre-ruminant calves (Van den Borne *et al.*, 2006), fat deposition was increased when nutrients were supplied in an asynchronous pattern.

The situation in ruminants is however, complicated by the effects of nutrient supply on both rumen and whole-body metabolism. A schematic pattern of energy and nitrogen (N) release in the rumen is presented in Figure 1. When N and energy are released at different rates (Figure 1A) then this would be considered as an asynchronous pattern. In grass silage, much of the N is rapidly available with a large proportion of the energy derived from the microbial digestion of fibre which would be considered as an asynchronous pattern. Supply of a rapidly degradable energy source such as ground barley or molasses would compliment the rapid release of N release in grass silage whilst a more slowly degraded N source such as formaldehyde treated soybean meal would compliment the slower energy release and result in a more synchronous pattern of nutrient release (Figure 1B). Many experiments that have been conducted in ruminants to evaluate the pattern of nutrient release in the rumen have relied upon matching the rate of degradation (or "c" co-efficient) of energy and protein rich feeds. A major disadvantage of this approach is that whilst the rate of degradation of the potentially degradable N in a feedstuff such as distillers grains is similar to that of the organic matter in sugarbeet pulp, the large soluble N fraction in distillers grains results in the combination of these feeds producing an asynchronous pattern of nutrient release.

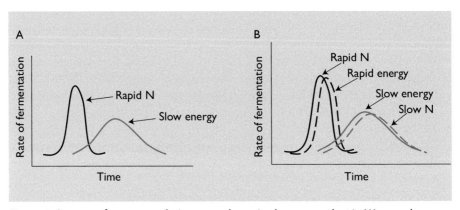

Figure 1. Pattern of energy and nitrogen release in the rumen that is (A) asynchronous and (B) synchronous (adapted from Johnson, 1976).

Ruminants posses the ability to recycle temporal patterns of excess N (either in the form of excess rumen degradable N or tissue amino acid N) to the rumen where it can be re-utilised and provide an ameliorating effect on ruminal ammonia concentrations. Consideration should also be given to the effects of pattern of supply of N substrates such as peptides and amino acids which have been shown in their own right to alter the efficiency of microbial protein synthesis. This lead Cabrita *et al.* (2006) in a recent review of the literature to conclude that the evidence for benefits to microbial protein synthesis (MPS) of matching nutrient supply to the rumen on a time-scale shorter than 24-h was contradictory. Therefore, from a viewpoint of practical rationing for ruminants, is moving to a time-scale shorter than 24-h just too complicated and impractical?

3. Infusion studies

A number of studies have investigated the pattern of nutrient supply to the rumen by infusing soluble energy and N solutions at differing rates. In some (e.g. Kim *et al.*, 1999b; Henning *et al.*, 1993), there was no effect of dietary synchrony on MPS. By contrast, Rooke *et al.* (1987) and Rooke and Armstrong (1989) reported that microbial efficiency was enhanced when glucose and casein supply were synchronised in cattle fed grass silage based diets. Similarly, Trevaskis *et al.* (2001) reported that the infusion of sucrose at 4 h post-ingestion of ryegrass pasture, when peak ammonia NH_3 concentrations were predicted, decreased peak NH_3 concentration and increased MPS by approximately 50%. Reasons for these discrepancies are unclear. A synchronous infusion of maltodextrose, which is considered more relevant to starch digestion in the rumen, resulted in a significant increase in microbial protein synthesis (Kim *et al.*, 1999a) and it was concluded that any beneficial effect may be limited to diets high in fermentable carbohydrate. Additionally, sugar sources can elicit different effects on microbial metabolism (Hussain and Miller, 1999) independent of their rate of release. It is also possible that the provision of soluble energy and nitrogen in a pulse dose may supply nutrients at a rate in excess of the microbial capacity to utilise them efficiently (Henning *et al.*, 1993)

4. Rumen metabolism *in vivo*

Several studies have been conducted to investigate the effect of matching the pattern of nutrient from different feed ingredients on ruminal metabolism (see review of Cabrita *et al.*, 2006). Mostly these have used different ingredients and been confounded by either not characterising the rate of nutrient release from all the dietary ingredients or have matched the rate of nutrient release from the principal ingredients but not considered the effects of the soluble fractions. This makes the evaluation of these studies little more than a catalogue of individual feed ingredients rather than an evaluation of the effect of pattern of nutrient release. Studies that have characterised nutrient release from all the dietary ingredients (by using, for example feedstuff in situ degradability co-efficients) or supplied the same ingredients at different times of the day, provide a more meaningful comparison. A summary of some of these is provided in Table 1. There is little consistent effect of altering the pattern of nutrient release within a 24-h period on microbial growth, with some studies indicating an increase,

Table 1. Effect of pattern of nutrient release in the rumen (synchronous, asynchronous or intermediate) determined either from in situ degradability of the feed ingredients or sequence of provision of feed ingredients, on microbial metabolism.

Dietary synchrony	Species	MPS gN/d	gN/kg OMADR[1]	Peak NH$_3$	Reference
Synch	Sheep	17.5	30.8	6	Sinclair *et al.*, 1993
Asynch		13.2	27.0	10	
Synch	Sheep	17.3	30.7	12	Sinclair *et al.*, 1995
Asynch		17.2	27.5	16	
Synch	Dairy cattle	328[a]	40.2	--	Henderson *et al.*, 1998
Int		361[b]	44.9	--	
Asynch		402[b]	53.8	--	
Synch	Beef cattle	70.1	26.5	9.5	Valkeners *et al.*, 2004
Int		73.5	30.7	11	
Asynch		72.6	28.5	12	
Synch	Beef cattle	73.5	28.3	30	Valkeners *et al.*, 2006
Int		72.1	25.1	34	
Asynch		72.5	21.8	26	
Synch	Beef cattle	52[a]	26.3	11	Chumpawadee *et al.*, 2006
Int-a		46[ab]	28.6	15	
Int-b		45[ab]	26.9	15	
Asnch		37[b]	25.0	23	

[1]OMADR: organic matter apparently degraded in the rumen.
[a-b]For MPS, within the same study, means in the same column with different superscripts differ $P<0.05$.

others no significant effect whilst some reported a decrease. The effects on peak ammonia concentrations appear more consistent, with more synchronous diets generally resulting in lower peaks and were associated with a reduced ruminal pH, perhaps indicative of increased microbial metabolism.

Ammonia produced in the rumen that is in excess of microbial requirement may be absorbed across the ruminal epithelium and converted to urea in the liver (Lobley *et al.*, 1995). The provision of an asynchronous nutrient regime in sheep has been shown to result in twice the re-cycling of N to the rumen (6.5 vs. 13.3 gN/d; Holder *et al.*, 1995) with the consequence of an ameliorating effect on ruminal ammonia concentrations, with mean values being higher in the lambs receiving the asynchronous regime. Increased amino-acid deamination associated with ammonia detoxification have been reported in growing lambs and isolated hepatocytes respectively (Lobley *et al.*, 1995; Mustvangwa *et al.*, 1997), resulting in a reduction in tissue amino acid supply which may have consequences on subsequent animal performance, particularly for higher producing animals. Additionally, elevated plasma ammonia levels are associated with a suppression of insulin (Choung and Chamberlain, 1995; Sinclair *et al.*, 2000b), which may be important due to the involvement of insulin in the regulation of growth, lactation and reproduction.

In theory, a reduction in ruminal ammonia concentrations may result in a decreased N excretion and therefore potentially reduce environmental pollution from more intensive ruminant production systems. The rumen ammonia pool is relatively small and turns over rapidly, with small changes in relative rates of production and removal resulting in rapid changes in rumen and subsequently plasma ammonia concentrations (Nolan, 1993). Most studies that have examined the effects of dietary synchrony on N output have, however, reported no effect on urinary-N excretion or total N balance (e.g. Witt *et al.*, 1999; Richardson *et al.*, 2003). Whilst ruminants have evolved the ability to re-cycle N in excess of microbial requirements to the rumen, a similar mechanism for energy conservation in the rumen is less apparent. Ruminal microbes can store energy in excess of requirements, although there is little evidence that this occurs to any great extent during short-term periods of energy excess (e.g. Sinclair *et al.*, 1993). Energy sufficient cultures of ruminal bacteria consume more energy and produce more heat than energy limited cultures (Russell, 1986). The advantages of this increased heat production are unclear but it is possible that ruminal bacteria use the proton motive force to transport ions in preparation for an eventual increase in growth rate. Certain micro-organisms such as *S. ruminatium* can also change from an acetate and propionate fermentation, where they yield

4 mols of ATP/mol hexose, to lactate, where 2 mol ATP/mol hexose is produced (Wallace, 1978). This results in ATP produced per unit time being maximised for microbial growth but a reduced feed efficiency (kg gain per unit feed intake) for the animal. In relation to whole-body metabolism, Richardson *et al.* (2003) reported that the efficiency of energy use (MJ retained/MJ fed) was significantly increased (Table 2) in lambs when fed the same ingredients in a synchronous compared with an asynchronous pattern, principally due to an increase in non-carcass and carcass fat levels and without an effect on overall growth rate. Few studies have demonstrated a similar effect in lactating animals, although the change-over design employed in many of these makes changes in body tissue composition difficult to detect.

Table 2. Effect of pattern of nutrient supply on the metabolism of growing lambs fed at a restricted level (from Richardson et al., 2003).

	Synch.	Int.	Asynch.	s.d.
Microbial growth (gN/d)	14.2	11.3	11.8	1.3
Peak plasma ammonia: M[1]	191[a]	145[b]	88[c]	4.6
N-balance g/g	0.165	0.157	0.160	0.016
Energy balance (MJ/MJ)	0.101[a]	0.100[a]	0.083[a]	0.009
Insulin: iu/ml	20	23	17	3.6

[1] Barley based diets.
[a-c] Within a row, means with different superscripts differ $P<0.05$.

5. Effects on animal performance

5.1. Growing animals

Witt *et al.* (1999b) reported a 22% increase in growth rate and a 23% improvement in food conversion when lambs were fed synchronous diets at a restricted level but few other studies have reported an improvement in performance either when animals have been fed at a restricted level or *ad libitum* (Table 3). For example, when the same diets used by Witt *et al.* (1999b) were offered *ad-libitum* to growing lambs (Witt *et al.*, 1999a), animals modified their pattern of intake with the consequence that growth rate was not affected, although food conversion was still improved by 8%. Beef heifers have also been shown on to

Table 3. Effect of pattern of nutrient supply to the rumen determined either from in situ degradability of the feed ingredients or sequence of provision of feed ingredients on the performance of growing lambs fed ad libitum or at a restricted level.

	Intake (kg DM/d)	Growth rate (g/d)	FCE (kg growth/kg feed intake)	Reference
Ad-lib intake				
Slow Synch	1.58	238	0.15[ab]	Witt *et al.*, 1999a
Slow Asynch	1.56	219	0.14[a]	
Fast Synch	1.47	259	0.18[c]	
Fast Asynch	1.67	272	0.16[bc]	
Synch	1.71	336	0.19	Sinclair *et al.*, unpublished data
Int	1.62	293	0.18	
Asynch	1.79	315	0.18	
Restricted intake				
Slow Synch	--	126[b]	0.13[b]	Witt *et al.*, 1999b
Slow Asynch	--	108[a]	0.11[a]	
Fast Synch	--	137[b]	0.15[b]	
Fast Asynch	--	107[a]	0.11[a]	
Slow Synch	1.05	182	0.18	Richardson *et al.*, 2003
Slow Int	1.05	185	0.18	
Slow Asynch	1.04	188	0.18	
Fast Synch	1.08	182	0.17	
Fast Int	1.06	193	0.19	
Fast Asynch	1.10	184	0.17	

[a-c] Within the same study, means in the same column with different superscripts differ $P<0.05$.

modify their pattern of intake when offered diets containing high levels of rapidly available N that were associated with elevated plasma ammonia concentrations (Sinclair *et al.*, 2000b). The mechanisms underlying this response are not fully understood but are thought to relate to the depletion of L-glutamate and L-aspartate which have both been shown to be powerful appetite stimulants, for ammonia metabolism in the brain. Therefore, ruminants appear to attempt to compensate by taking a greater number of smaller meals which may not in itself decrease total DM intake unless the animal is restricted in intake either by time and/or feeding conditions.

5.2. Dairy animals

Cabrita *et al.* (2006) reviewed the effects of dietary synchrony on dairy cow performance and reported no consistent effect on milk performance. Again, dietary synchrony was poorly defined in many of the studies and findings were contradictory, with ingredient effects appearing as important as the pattern of nutrient release. Additionally, secondary effects of nutrient release, on ruminal pH in particular, often masked any direct effects of dietary synchrony on animal performance. Several studies also employed change-over designs to evaluate effects on milk performance, making effects of pattern of nutrient supply on the partioning to body tissue difficult to determine. It can be concluded however, that beneficial effects of synchronising nutrient supply to the rumen may be limited to situations of discrete meal provision or more extreme patterns of nutrient release.

5.3. Effects of nitrogen intake on fertility

It is generally recognised that in ruminants high protein diets are associated with a reduction in reproductive performance, particularly conception rate to first service. There are several means by which dietary protein may elicit this response, including increasing DM intake and enhancing body fat mobilisation, particularly when undegradable protein levels have been increased (Laven and Drew, 1999). Negative impacts have also been reported with increasing levels of rumen degradable protein leading to increased plasma concentrations of urea and ammonia (Butler, 1998) with most attention being focussed on urea and its relationship to fertility, as urea is easily measured and relatively stable. The influence of urea on fertility has traditionally been considered to occur via direct effects on the modification of oviductal and uterine environments through altering early pre-implantation embryo development. Despite this, there is often a poor relationship between milk urea-N and indices of fertility (Figure 2). This suggests that urea *per se* may not be the causal factor, or that problems of fertility are dependent on the nature and/or pattern of ammonia release in the rumen and subsequent metabolism. The liver appears to have a maximal ability to metabolise ammonia of around 0.68 mol/h (Di Marco *et al.*, 1998), although studies in sheep and cattle have shown that liver metabolism can adapt to high levels. For example, animals that had previously been exposed to high-N diets were better able to maintain jugular ammonia levels when

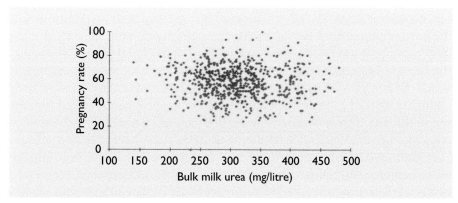

Figure 2. Relationship between bulk milk tank urea concentrations and pregnancy rate in dairy cows (from Cottrill et al.*, 2002).*

infused or fed additional nitrogen (Morris and Payne, 1970). Nevertheless, a supply to the liver above approximately 12 g of ammonia per hour may exceed the livers capacity to convert it into urea. Lactating cows in negative energy balance also seem to be less tolerant of high levels of plasma ammonia and/or urea than non-lactating cows or when in positive energy balance.

Oocytes recovered from cattle that had been fed diets that resulted in high plasma ammonia/urea concentrations have been shown to have a significantly poorer capacity to develop when cultured in vitro, with less than half the number reaching the blastocyst stage (Sinclair *et al.*, 2000a). More recently, the effect of the maternal diet several weeks prior to fertilisation has been shown to have a major influence on subsequent fertility rates. Studies where ovine embryo donors were fed high N diets and the embryos subsequently implanted in recipients receiving a standard diet, resulted in accelerated early embryo development, reduced pregnancy rates (16% vs. 39% in control) and significantly altered foetal development and gene expression (Powell *et al.*, 2006). Therefore, maternal N nutrition both at and several weeks prior to ova release can have a major influence on subsequent fertility rates.

6. Non-protein nitrogen (NPN)

6.1. Sources of NPN

Sources of non-protein nitrogen (NPN) are attractive because of their low cost relative to vegetable proteins. Whereas protozoa do not use ammonia, bacteria can, depending on the diet, derive between 40 to 95% of their N from ammonia, the balance being from peptides and amino acids (Nolan, 1993). Urea is the most commonly used NPN source and is rapidly hydrolysed to ammonia within the rumen. There has been a reluctance to feed urea to lactating ruminants which can be related to its potential effects on intake and fertility that have been outlined above, or even death by toxicity if it is inappropriately mixed with the rest of the diet. This has lead to the development of several methods aimed at reducing the rapid rate of N release from urea. These have included calcium chloride bound urea, biuret, isobutylidene diurea, acetyl urea, tung- and linseed-oil-coated urea, formaldehyde treated urea and natural zeolite. Commercial uptake of these slower release NPN sources is dependent on their cost relative to urea and vegetable protein sources, and their effects on microbial growth and animal performance. Additionally, for maximal utilisation of certain slow release N sources such as biuret, a variable but required period of adaptation has been reported (Johnson and Clemens, 1973). The length of this adaptation period is dependent on the starch and protein content of the diet but can be as high as 42 days (Johnson and Clemens, 1973), and therefore limits the use of biuret in many practical feeding situations. Other protection techniques may bind the urea too tightly rendering it unavailable in the rumen for hydrolysis whilst others have been demonstrated to have little effect on reducing the rate of urea release in the rumen.

6.2. Non-Protein Nitrogen sources and microbial metabolism in vitro

Recently, a fatty acid blended urea source (Optigen®) has been developed, and a summary of studies that have evaluated its effect *in vitro* are presented in Table 4. Replacement of pre-formed protein sources with Optigen® appear to result in a similar amount and efficiency of microbial protein synthesis but enhanced values were recorded when Optigen® replaced feed grade urea. Interestingly, in the study of Sinclair *et al.* (unpublished results), fibre digestibility was also increased when Optigen replaced soyabean and rapeseed meal, despite ammonia-N values being excess in both treatments (Table 5). Fewer studies have been conducted on the effects of slow release protein sources on animal performance, with the majority of those that have focussing on high fibre diets fed to low producing animals. For example, Currier *et al.* (2004) reported that NPN supplementation increased suckler cow weight and condition score at calving, even when fed every other day but there was little difference between urea and biuret. In higher yielding dairy cows, Galo *et al.* (2003) reported a similar level of milk production when a polymer coated urea source replaced a combination of urea and pre-formed protein sources whilst a liquid calcium

Table 4. Studies on the effect of replacing vegetable protein sources or urea with a slow release urea source (Optigen) on microbial metabolism in vitro.

	pH	Ammonia-N (mg/dL)	DM digested (%)	Microbial growth (gN/d)	Microbial efficiency gN/kg OMADR[1]	Reference[2]
Urea-low	6.5	4.0	42	0.21	17	1
Optigen-low	6.4	6.5	44	0.27	20	
Urea-high	6.4	3.8	43	0.15	13	1
Optigen-high	6.4	4.2	43	0.24	18	
Control	6.4	4.1	62	0.35	25	2
Optigen	6.4	4.4	63	0.36	25	
Control	6.4	4.2	60	0.32	23	3
Optigen	6.5	4.4	61	0.32	23	
Urea	6.4	6.2	62	0.34	24	4
Optigen	6.4	7.3	63	0.35	25	
Control	6.4	3.0	59	0.29	22	5
Optigen	6.5	3.2	58	0.29	22	
Control	6.2	20	37	0.35	34	6
Optigen	6.2	23	49	0.39	30	

[1] OMADR: organic matter apparently degraded in the rumen.
[2] References: (1) Tikofsky and Harrison, 2006; (2) Harrison *et al.*, 2007a; (3) Harrison *et al.*, 2007b; (4) Harrison *et al.*, 2007c; (5) Harrison *et al.*, 2007d; (6) Sinclair and Huntington, unpublished results.

Table 5. Ruminal volatile fatty acid concentration and digestibility in continuous culture vessels given diets containing soybean meal and rapeseed meal (Control) or with partial replacement of the soybean meal and rapeseed meal with a slow release urea source (Optigen®).

	Control	Optigen	s.d.	P-value
Mean fluid total VFA (mmol/l)	80.3	86.9	2.92	0.151
Mean fluid acetate to propionate ratio	2.5	2.5	0.23	0.968
Digestibility (g/g)				
Organic matter	0.37	0.49	0.039	0.103
Fibre	0.45	0.57	0.028	0.046

chloride-bound urea resulted in a similar level of milk production to cows fed soyabean meal but at a reduced DM intake with a correspondingly improved feed efficiency (Golombeski *et al.,* 2006; Table 6). By contrast, in beef animals, the addition of zeolite had little benefit over urea alone as a NPN source on animal performance, although fibre digestibility was higher (Sadeghi and Shawrang, 2006), whilst there was little benefit to replacing feed urea with a

Table 6. Some recent studies in cattle comparing vegetable protein sources with urea or slow-release urea sources on animal performance.

	Control	Slow release urea	Urea	Reference
Dairy cows				
DM intake (kg/d)	23.6	23.6	-	[1]Galo *et al.,* 2003
Milk yield (kg/d)	35.6	34.8	-	
Fat (g/kg)	38	36	-	
Protein (g/kg)	31	31	-	
DM intake (kg/d)	21.3[a]	19.9[b]	-	[2]Golombeski *et al.,* 2006
Milk yield (kg/d)	26.1	26.2	-	
Fat (g/kg)	43	44	-	
Protein (g/kg)	37	37	-	
Beef cattle				
DM intake (kg/d)	7.9	7.8	7.8	[3]Sadeghi and Shawrang, 2006
Liveweight gain (kg/d)	1.4[a]	1.2[b]	1.3[c]	
kg gain/kg feed	0.18[a]	0.16[b]	0.16[b]	
DM intake (kg/d)	-	9.0	9.4	[4]Tedeschi *et al.,* 2002
Liveweight gain (kg/d)	-	1.48	1.64	
kg gain/kg feed	-	0.16	0.17	
DM intake (kg/d)	-	9.4	9.3	[4]Tedeschi *et al.,* 2002
Liveweight gain (kg/d)	-	1.42	1.65	
kg gain/kg feed	-	0.16	0.17	

[1] Slow release urea = polymer coated urea.
[2] Slow release urea = liquid calcium chloride-bound urea.
[3] Slow release urea = urea added to zeolite.
[4] Slow release urea = polymer coated.

slow release urea product in the study of Tedeschi *et al.,* (2002). Benefits to slow release urea sources will be dependent on a number of factors, including the ability of the animal to metabolise excess N released in the rumen and further studies are required to evaluate the effect of these slower released urea sources as replacements for urea and vegetable protein sources particularly in the diet of high yielding dairy cows.

6.3. Requirement for pre-formed amino acids

Despite the potential financial advantages of NPN sources, benefits to supplying pre-formed amino acids and peptides on microbial growth and efficiency are well established. It has been shown, for example, that the yield of non-fibryloytic bacteria is improved by as much as 19% as the ratio of peptides to non-structural carbohydrates plus peptides increases from 0-14% (Russell and Sniffen, 1984), and under conditions of large amounts of readily fermentable carbohydrates it has been suggested that there may be a limitation of peptides for fibrylolytic microbial growth (Hoover, 1986). It is therefore important that when pre-formed protein sources are replaced with NPN that sufficient rumen available amino acids and peptides are provided from a variety of feed ingredients.

7. Implications

There are a number of controlled studies that indicate that matching nutrient supply to the rumen on a time-scale shorter than 24-h can improve microbial metabolism. The difficulty in accurately predicting the rate of nutrient release from different batches of feeds and the influence of altered meal patterns in association with the ameliorating effect of re-cycling N in excess of microbial requirements to the rumen negative, currently make it difficult to quantitatively implement such an approach. Future feed supplies for cattle are likely to see large fluctuations in raw material prices and the current up-turn in the cost of protein feeds make NPN sources more attractive. The evidence from the literature indicates that attempts to capture the rapidly released N from feed grade urea with very rapidly released energy sources such as sugars are unlikely to be very successful. Non-protein nitrogen sources offer the opportunity of replacing around 50% of the rumen degradable protein supply, providing the diet supplies sufficient amino acids to meet microbial and animal requirements. Within NPN sources, there may be a benefit to slow release urea sources in situations where the animals ability to metabolise and re-cycle urea is exceeded. There are a number of slow-release urea sources with varying degrees of efficacy and the effects of their incorporation into the diet of high yielding dairy cows on performance warrants further research.

References

Butler, W.R., 1998. Review: Effect of protein nutrition on ovarian and uterine physiology in dairy cattle. *Journal of Dairy Science* **81:**2533-2539.

Cabrita, A.R.J., R.J. Dewhurst, J.M.F. Abreu and A.J.M. Froncesca, 2006. Evaluation of the effects of synchronising the availability of N and energy on rumen function and production responses of dairy cows – a review. *Animal Research* **55:**1-24.

Choung, J.J. and D.G. Chamberlain, 1995. Effects of intraruminal infusion of propionate on the concentrations of ammonia and insulin in peripheral blood of cows receiving an intraruminal infusion of urea. *Journal of Dairy Research* **62:**549-557.

Chumpawadee, S., K. Sommart, T. Vongpralub and V. Pattarajinda, 2006. Effects of synchronising the rate of dietary energy and nitrogen release on ruminal fermentation, microbial protein synthesis, blood urea nitrogen and nutrient utlisation in beef cattle. *Asian-Australian journal of Animal Sciences* **19:**181-188.

Cottrill, B., H.J. Biggadike, C. Collins, and R.A. Laven, 2002. Relationship between milk urea concentration and the fertility of dairy cows. *The Veterinary Record* **151:**413-416.

Currier, T.A., D.W. Bohnert, S.J. Falck, and S.J. Bartle, 2004. Daily and alternate day supplementation of urea or biuret to ruminants consuming low-quality forage: 1. Effects on cow performance and the efficiency of nitrogen use in wethers. *Journal of Animal Science* **82:** 1508-1517.

Di Marco, O.N., P. Castiñeiras and M.S. Aello, 1998. Ruminal ammonia concentration and energy expenditure of cattle estimated by the carbon dioxide entry rate technique. *Animal Science* **67:**435-443.

Galo, E., S.M. Emanuele, C.J. Sniffen, J.H. White and J.R. Knapp, 2003. Effects of a polymer coated urea product on nitrogen metabolism in lactating Holstein dairy cattle. *Journal of Dairy Science* **86:**2154-2162.

Golombeski, G.L., K.F. Kalscheur, A.R. Hippen and D.J. Schingoethe, 2006. Slow-release urea and highly fermentable sugars in diets fed to lactating dairy cows. *Journal of Dairy Science* **89:**4395-4403.

Harrison, G.A., J.M. Tricarico, M.D. Meyer and K.A. Dawson, 2007a. Effects of Optigen®II on fermentation, digestion, and N partioning in rumen-simulating fermenters fed diets with distillers dried grains. In: *Nutritional Biotechnology in the Feed and Food Industries: Proceedings of Alltech's 23rd Annual Symposium* (Eds. T.P. Lyons., K.A. Jacques and J.M. Hower). Abstract No. 306.

Harrison, G.A., J.M. Tricarico, M.D. Meyer and K.A. Dawson, 2007b. Effects of Optigen®II on fermentation, digestion, and N partioning in rumen-simulating fermenters. In: *Nutritional Biotechnology in the Feed and Food Industries: Proceedings of Alltech's 23rd Annual Symposium* (Eds. T.P. Lyons., K.A. Jacques and J.M. Hower). Abstract No. 307.

Harrison, G.A., J.M. Tricarico, M.D. Meyer and K.A. Dawson, 2007c. Effects of urea or Optigen®II on fermentation, digestion, and N flow in rumen-simulating fermenters. In: *Nutritional Biotechnology in the Feed and Food Industries: Proceedings of Alltech's 23rd Annual Symposium* (Eds. T.P. Lyons., K.A. Jacques and J.M. Hower. Abstract No. 308.

Harrison, G.A., J.M. Tricarico, M.D. Meyer and K.A. Dawson, 2007d. Influence of degradable protein and fermentable carbohydrate on Optigen®II effects on fermentation, digestion, and N flow in rumen-simulating fermentors. In: *Nutritional Biotechnology in the Feed and Food Industries: Proceedings of Alltech's 23rd Annual Symposium* (Eds. T.P. Lyons., K.A. Jacques and J.M. Hower). Abstract No. 351.

Henderson, A.R., P.C. Garnsworthy, J.R. Newbold and P.J. Buttery, 1998. The effect of asynchronous diets on the function of the rumen in the lactating dairy cow. *Proc. Winter Meeting of the Br. Soc. Anim. Sci.*, p19.

Henning, P.H., D.G. Steyn and H.H. Meissner, 1993. Effect of synchronisation of energy and nitrogen on rumen characteristics and microbial growth. *Journal of Animal Science* **71**:2516-2528.

Holder, P., P.J. Buttery and P.C. Garnsworthy, 1995. The effect of dietary asynchrony on rumen nitrogen recycling in sheep. *Proc. Winter Meeting of the Br. Soc. Anim. Sci.*, p70.

Hoover, W.H., 1986. Chemical factors involved in ruminal fibre degradation. *Journal of Dairy Science* **69**:2755-2766.

Hussain, A. and E.L. Miller, 1999. Effect of supplementation of sucrose and lactose with sodium bicarbonate on rumen metabolism and microbial protein synthesis in sheep. *Proc. Winter Meeting of the Br. Soc. Anim. Sci.*, pp. 28.

Johnson, R.R., 1976. Influence of carbohydrate solubility on non-protein nitrogen utilization in the ruminant. *Journal of Animal Science* **43**:184-191.

Johnson, R.R. and E.T. Clemens, 1973. Adaptation of rumen microorganisms to biuret as an NPN supplement to low quality roughage rations for cattle and sheep. *Journal of Nutrition* **103**:494-502.

Kim, K.H., J.J. Choung and D.G. Chamberlain, 1999a. Effects of varying the degree of synchrony of energy and nitrogen release in the rumen on the synthesis of microbial protein in lactating dairy cows consuming a diet of grass silage and a cereal-based concentrate. *Journal of the Science of Food and Agriculture* **79**:1441-1447.

Kim, K.H., Y G. Oh, J.J. Choung and D.G. Chamberlain, 1999b. Effects of varying degrees of synchrony of energy and nitrogen release in the rumen on the synthesis of microbial protein in cattle consuming grass silage. *Journal of the Science of Food and Agriculture* **79**:833-838.

Laven, R.A. and S.B. Drew, 1999. Dietary protein and the reproductive performance of cows. *Veterinary Record* **145**:687-695.

Lobley, G.E., A. Connell, M.A. Lomax, D.S. Brown, E. Milne, A.G. Calder and D.A.H. Farningham, 1995. Hepatic detoxification of ammonia in the ovine liver: Possible consequences for amino acid metabolism. *British Journal of Nutrition* **73**:677-685.

Morris, J.G. and E. Payne, 1970. Ammonia and urea toxicoses in sheep and their relation to dietary nitrogen intake. *Journal of Agricultural Science* **74**:259-271.

Mustvangwa, T., J.G. Buchanan-Smith and B.W. McBride, 1997. Effects of ruminally degradable nitrogen intake and *in vitro* addition of ammonia and propionate on the metabolic fate of L-[^{15}N]alanine in isolated sheep hepatocytes. *Journal of Animal Science* **75**:1149-1159.

Nolan, J.V., 1993. Nitrogen kinetics. In: (J.M. Forbes and J. France, Eds.) Quantitative aspects of ruminant digestion and metabolism. CAN international, Oxon, UK.

Powell, K., J.A. Rooke, T.G. McEvoy, C.J. Ashworth, J.J. Robinson, I. Wilmut, L.E. Young and K.D. Sinclair, 2006. Zygote donor nitrogen metabolism and in vitro embryo culture perturbs in utero development and IGF2R expression in ovine fetal cultures. *Theriogenology* **66**:1901-1912.

Richardson, J.M., R.G. Wilkinson and L.A. Sinclair, 2003. Synchrony of nutrient supply to the rumen and dietary energy source and their effects on the growth and metabolism of lambs. *Journal of Animal Science* **81**:1332-1347.

Rooke, J.A. and D.G. Armstrong, 1989. The importance of the form of nitrogen on microbial protein synthesis in the rumen of cattle receiving grass silage and continuous intrarumen infusions of sucrose. *British Journal of Nutrition* **61**:113-121.

Rooke, J.A., N.H. Lee, and D.C. Armstrong, 1987. The effects of intraruminal infusions of urea, casein, glucose syrup and a mixture of casein and glucose syrup on nitrogen digestion in the rumen of cattle receiving grass-silage diets. *British Journal of Nutrition* **57**:89-98.

Russell, J.B., 1986. Heat production by ruminal bacteria in continuous culture and its relationship to maintenance energy. *Journal of Bacteriology* **168**:694-701.

Russell, J.B., 1998. Strategies that ruminal bacteria use to handle excess carbohydrate. *Journal of Animal Science* **76**:1955-1963.

Russell, J.B. and C.J. Sniffen, 1984. Effect of carbon-4 and carbon-5 volatile fatty acids on growth of mixed rumen bacteria *in vitro*. *Journal of Dairy Science* **67**:987-994.

Sadeghi, A.A. and P. Shawrang, 2006. The effect of natural zeolite on nutrient digestibility, carcass traits and performance of Holstein steers given a diet containing urea. *Animal Science* **82**:163-167.

Sinclair, K.D., M. Kuran, F.E. Gebbie, R. Webb and T.G. McEvoy, 2000a. Nitrogen metabolism and fertility in cattle. II. Development of oocytes recovered from heifers offered diets differing in their rate of nitrogen release in the rumen. *Journal of Animal Science* **78**:2670-2680.

Sinclair, K.D., L.A. Sinclair and J.J. Robinson, 2000b. Nitrogen metabolism and fertility in cattle: 1. Adaptive changes in intake and metabolism to diets differing in their rate of energy and nitrogen release in the rumen. *Journal of Animal Science* **78**:2659-2669.

Sinclair, L.A., P.C. Garnsworthy, J.R. Newbold and P.J. Buttery, 1993. Effect of synchronising the rate of dietary energy and nitrogen release on rumen fermentation and microbial protein synthesis in sheep. *Journal of Agricultural Science* **120**:251-263.

Sinclair, L.A., P.C. Garnsworthy, J.R. Newbold and P.J. Buttery, 1995. Effect of synchronising the rate of dietary energy and nitrogen release in diets with a similar carbohydrate composition on rumen fermentation and microbial protein synthesis in sheep. *Journal of Agricultural Science* **124**:463-472.

Tedeschi, L.O., M.J. Baker, D.J. Ketchen and D.G. Fox, 2002. Performance of growing and finishing cattle supplemented with a slow-release urea product and urea. *Canadian Journal of Animal Science* **82**:567-573.

Tikofsky, J. and G.A. Harrison, 2006. Optigen®II: improving the efficiency of nitrogen utilization in the dairy cow. In: *Nutritional Biotechnology in the Feed and Food Industries: Proceedings of Alltech's 22nd Annual Symposium* (Eds. T.P. Lyons., K.A. Jacques and J.M. Hower), pp. 379-386.

Trevaskis, L.M., W.J. Fulkerson and J.M. Gooden, 2001. Provision of certain carbohydrate-based supplements to pasture-fed sheep, as well as time of harvesting of the pasture, influences pH, ammonia concentration and microbial protein synthesis in the rumen. *Australian Journal of Experimental Agriculture* **41**:21-27.

Valkeners, D., A. Théwis, S. Amant and Y. Beckers, 2006. Effect of various levels of imbalance between energy and nitrogen release in the rumen on microbial protein synthesis and nitrogen metabolism in growing double-muscled Belgian Blue bulls fed a corn silage-based diet. *Journal of Animal Science* **84**:877-885.

Valkeners, D., A. Théwis, F. Piron and Y. Beckers, 2004. Effect of imbalance between energy and nitrogen supplies on microbial protein synthesis and nitrogen metabolism in growing double-muscled Belgian Blue bulls. *Journal of Animal Science* **82**:1818-1825.

Van den Borne, J.J.G.C., J.W. Schrama, M.J.W. Heetkamp, M.W.A. Verstegen and W.J.J. Gerrits, 2007. Synchronising the availability of amino acids and glucose increases protein retention in pigs. *Animal* **1**:666-674.

Van den Borne, J.J.G.C., M.W.A. Verstegen, S.J.J. Alferink, R.M.M. Giebels and W.J.J. Gerrits, 2006. Effects of feeding frequency and feeding level on nutrient utilization in heavy preruminant calves. *Journal of Dairy Science* **89**:3578-3586.

Wallace, R.J., 1978. Control of lactate production by *Selenomonas ruminatum*: homotrophic activation of lactate dehydrogenase by pyruvate. *Journal of General Microbiology* **107**:45-52.

Witt, M.W., L.A. Sinclair, R.G. Wilkinson and P.J. Buttery, 1999a. The effects of synchronizing the rate of dietary energy and nitrogen supply to the rumen on the production and metabolism of sheep: food characterisation and growth and metabolism of ewe lambs given food ad-libitum. *Animal Science* **69**:223-235.

Witt, M.W., L.A. Sinclair, R.G. Wilkinson and P.J. Buttery, 1999b. The effects of synchronizing the rate of dietary energy and nitrogen supply to the rumen on the metabolism and growth of ram lambs given food at a restricted level. *Animal Science* **69**:627-636.

Calf gut health: practical considerations

D. O'Rourke
Ortec Consultancy, Kent, United Kingdom

1. Introduction

> *"Mortality has not really improved over the last 30 years and it is quite surprising that the accepted level of live calves up to three months of age is 90 for every 100 cows in calf up to seven months of gestation."*

> (Andrews, 1999)

What is important is that these calves are the replacements and future lifeblood of our herds. It is amazing that we accept a mortality rate of 10% after the investment we have put into getting the cow in calf, feeding her during the gestation period and feeding the newborn calf.

2. Incidence and causes of mortality

A recent study (Bricknell *et al.*, 2007; Table 1) confirmed that eight years later there has been little change. They found that the average perinatal mortality (stillbirths and mortality of male and female calves during the first 24 hours of life) was 8.1% and neonatal mortality (female calves born alive that die between 24 hours and 28 days) averaged at 3.4%. There was also a trend for neonatal mortality to increase with increasing herd size (2.2% <200 cows, 4.9% >400cows) and more calves died in autumn winter (6.2%) compared with spring and summer (2.6%) (P<0.05).

Table 1. Incidence of calf mortality on dairy farms in southern England (Bricknell et al., *2007).*

	% (Min – Max)	Previous estimate (%)
Perinatal mortality	8.1 (3–14)	7.8[1]
Neonatal mortality	3.4 (0–13)	10[2]

[1] Esslemont and Kossaibati, 1996.
[2] Mellor and Stafford, 2004.

Calf scours are the most common cause of mortality and morbidity (animals affected) during the first few weeks of life. In a survey of dairy calf management, mortality and morbidity in Ontario Holstein herds, 4% of liveborn heifer calves died, 20% of calves were treated for scours and 15% were treated for pneumonia before the age of weaning (Waltner-Toews *et al.*, 1986a). The incidence of neonatal calf diarrhoea reported by farmers in two practice areas in Bavaria was 15.4 and 28.4%, respectively (Katikaridis, 2000; Biewer, 2001). In a more recent study (Girnus, 2004), 205 calves in 25 farms were examined daily for the first 14 days of their lives by a veterinarian. Incidence of diarrhoea (defined as loose faeces flowing through the spread fingers of a gloved hand on at least one examination) was 47.8% with a range from 0 to over 90% across farms.

Causes of calf scour are bacterial (*E. coli*, *Salmonella* spp.), viral (Rotavirus, Coronavirus) and protozoal (*Cryptosporidium parvum*, *Eimeria* spp.) in nature. However, the challenge from these pathogens does not occur at the same time (Figure 1). A survey in Ontario Holstein herds found that the percentage of calves treated for scours peaked during the second week of life at 8.1%, then declined sharply, approaching zero by about six weeks (Waltner-Toews *et al.*, 1986b).

A comprehensive investigation on calf enteritis in the United Kingdom involving both husbandry and pathogens, carried out by the National Animal

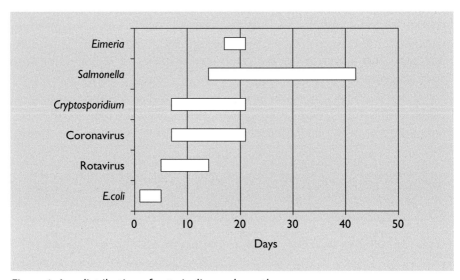

Figure 1. Age distribution of enteric disease by pathogen.

Disease Information Service (NADIS), compared 256 scour outbreaks with 79 unaffected control farms (Andrews, 2000). Infectious organisms were found in 190 (74.2%) of outbreaks and 62 (24.2%) of these were mixed infections. The organisms isolated from these outbreaks are shown in Table 2. In 66 (25.8%) of outbreaks no organism was found indicating that other factors (e.g. nutrition) can cause calf scours.

During the period 2002 to 2006 approximately 4% of submissions to the Veterinary Laboratories Agency (VLA) were found to be positive for Coronavirus. The report indicated that hypogammaglobulinaemia, as a consequence of inadequate colostral antibody absorption, continued to be a principal precipitating cause of neonatal calf scours (VLA, 2006).

The main issue that has to be addressed in a scouring animal is dehydration and acidosis. In the NADIS survey, treatment involved the use of fluid therapy in 86% of outbreaks and antimicrobials in 86% also. In over two-thirds of cases (71%) both forms of therapy were needed. Each case of calf scour is estimated to have an overall cost of €185 per sick animal (Milk Development Council, 1998).

Treatment for scours can have a long term effect on the survival of the calf in the herd and age at first calving. Heifers with a calfhood history of being treated for scours were 2.5 times more likely to be sold than other calves. Heifers which had been treated for scours were 2.9 times more likely to calve after 900 days of age than other heifers (Waltner-Toews *et al.*, 1986c).

Table 2. The number of outbreaks from which each infectious organism was isolated (Andrews, 2004).

Organism	Number of outbreaks	% of outbreaks
Rotavirus	101	37.3
Cryptosporidia	80	29.5
Coccidia	30	11.1
Coronavirus	26	9.6
E. coli K99	25	9.2
Salmonella spp.	9	3.3
Total	271	100

Prevention and control of calf scours starts well before the calf is born. If the animal needs treating with antimicrobials, then you may have already lost the battle.

3. Lack of immunity

The newborn calf is devoid of immunity as antibodies cannot cross the placental wall from the dam to the foetus. Therefore, the calf is totally reliant on the antibodies it receives by consuming adequate quantities of colostrum within the first few hours after birth. These antibodies help to protect the calf against infection during the first few weeks of life (period of high risk) until their immune systems are fully functional (Figure 2).

One should always remember that the dam's colostrum will contain antibodies (immunoglobulins) to the pathogens that it has been exposed to. This is important where heifers are reared on another farm and brought to the home farm just prior to calving, as they may not have immunity against the endemic pathogens on the home farm.

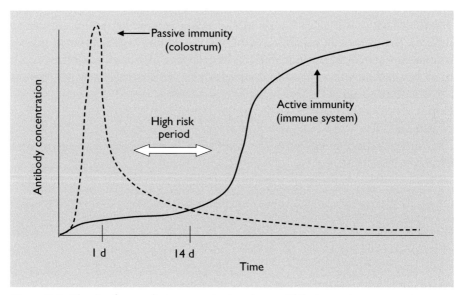

Figure 2. Antibodies from colostrum protects calves until their own immune systems are fully functional (Heinrichs and Jones, 2003).

4. Management of calf scours

Management of calf scours starts well before the calf is born. There are three phases of the cycle – prior to calving, at calving and post calving.

4.1. Pre-calving

Feeding should always be aimed at meeting the nutritional requirements of the dairy cow during the dry period. During the transition period the dairy cow can experience negative energy balance which can have a major impact on the immune response. Cows should be fit but not fat at calving (body condition score 3, on a scale of 1-5). Gross overfeeding may lead to calving difficulties including adverse effects on the viability of the calf. Underfeeding can also affect the cow and foetus.

Vaccines are available to stimulate immunity against particular pathogens (e.g. *E. coli*, Rotavirus, Coronavirus), thereby, helping to provide further protection to the calf via the colostrum.

Franklin *et al.* (2005) found that cows supplemented with 10g/day of mannan-oligosaccharides (Bio-Mos®, Alltech Inc.) for the last 4 weeks of the dry period had greater rotavirus titres at calving in response to vaccination against rotavirus. Their calves tended to have greater rotavirus titres at 24 hours after birth compared with control cows and their calves (Figure 3). Anything that can be done to increase the level of antibodies in colostrum and, thereby, help the calf withstand the challenges in the first few days of life should be considered. This is far better than stressing the calf with vaccination in early life. Under most circumstances, calves that receive adequate colostrum from vaccinated dams have little or no need to be vaccinated before weaning.

4.2. At calving

A recent study (Carrier *et al.*, 2005) investigated factors that affect the number of stillborn calves in a commercial herd. Twin births were not included in the study, and a total of 495 calvings were analysed. Results indicated that cows requiring slight assistance with calving (calving difficulty score of 2 on a 5-point scale) were 2.9 times more likely to have a stillborn calf than cows that delivered unassisted. Stillbirth was 46 times more likely for cows with a calving difficulty score of 3 or more compared to cows that did not require assistance. Heifers were 5.2 times more likely than cows to have a stillborn calf. Although

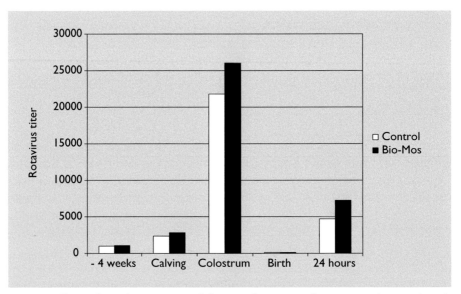

Figure 3. Effects of supplementation of close-up dry cows with mannan-oligosaccharide (Bio-Mos®) on immune function and transfer of passive immunity to their calves.

these results were from a single herd they demonstrate that any type of assisted delivery increases risk to the calf. Another finding was that cows moved from free stalls to an individual calving pen when the water bag or feet/head were showing were 2.5 times less likely to have stillborn calves than cows moved while only mucus or blood were showing.

One of the key issues at calving is colostrum and there are a number of factors that can affect the immunoglobulin concentration. Cows that produce a large quantity of colostrum (greater than 8 kg at first milking) often produce lower concentration of immunoglobulin which is probably due to dilution. The immune status of the dam will affect the immunoglobulin level and this depends on pathogen exposure and vaccination status. A 3 to 4-week dry period is needed to allow antibodies from the blood to be concentrated in colostrum. Dry cow nutrition is important as cows fed too little protein or energy tend to produce lower quality colostrum than cows fed adequately. Two year-old cows often have the poorest colostrum quality and this is probably related to a lack of exposure to pathogens. Leaking milk prepartum or milking before calving can both reduce antibody levels by colostrum removal or by dilution. Jerseys tend to have the highest levels of antibodies averaging 66 g/l of IgG with a range of 28 to 115 g/l whereas Holsteins are the lowest typically averaging 48.2 g/l of IgG with a range of 20 to 100 g/l of colostrum.

High quality colostrum has an IgG concentration >50 g/l. However, the quality of colostrum from different cows and farms can be highly variable. In a recent study in the US, colostrum IgG averaged 76 mg/ml with a range from 9 to 186 g/l for individual cows (Godden, 2007).

The calf's ability to absorb antibodies declines rapidly over the first 24 hours. Therefore, the first 24 hours are essential and the calf should receive 4 litres in the first 4 hours which is equivalent to 20 minutes of hard sucking. Twenty five percent of calves left alone do not nurse within 8 hours and 10 to 25% do not get enough colostrum.

One of the key issues with colostrum is hygiene. It is imperative that it is as clean as possible, as the calf's gut is sterile and pathogenic bacteria in colostrum could cause diseases such as scours or septicaemia. Also, the bacteria may prevent antibodies from being absorbed and reaching the calf's blood (James *et al.*, 1981). It is recommended that fresh colostrum fed to calves should contain less than 100,000 cfu/ml total bacteria count (TBC) and less than 10,000 cfu/ml total coliform count (McGuirk and Collins, 2004). Studies have indicated that average levels of bacteria in colostrum fed to calves are significantly higher than this. In a recent study in Wisconsin dairy herds, 82% of samples tested exceeded the upper limit of 100,000 cfu/ml TBC (Poulson *et al.*, 2002).

The importance of colostrum feeding was shown in a recent study that evaluated the effects of feeding two different volumes of colostrum immediately after birth on the growth and subsequent lactational performance of heifers (Faber *et al.*, 2005). Heifer calves were fed either 2 l or 4 l of high quality colostrum within the first hour of birth. Second and subsequent milk feedings were identical for both groups. Each animal received some colostrum daily for 14 days. Veterinary costs were approximately doubled for heifers fed 2 l of colostrum compared with heifers fed 4 l of colostrum. Those animals fed 4 l gained significantly greater daily body weight compared with herdmates fed 2 l (1.03 vs. 0.80 kg; P<0.001). Animals fed 4 l of colostrum at birth produced significantly more milk compared with those fed 2 l (9,907 and 11,294 kg versus 8,952 and 9,642 kg at first and second lactations, respectively) when lactation records were adjusted as 305-d mature equivalent. Overall, feeding a greater volume of quality colostrum resulted in an increase of 1349 kg of actual milk produced per cow in the second lactation. This would result in a direct economic return to the producer of approximately €160 per cow.

Other sources of infection can be bedding, water and feeding utensils. Therefore, it is imperative that good hygiene practices are in place to ensure a comfortable clean dry environment for the calving cow as well as the calf during the first few months of life. Procedures should be in place to ensure that colostrum and feeding utensils are not contaminated before us. There is a simple equation that pathogen load plus degree of protection from colostrum equals calf health.

4.3. Post-calving

It is important that management practices are put in place to ensure the calf receives sufficient colostrum. However, a recent study to determine the levels of passive immunity, as measured by serum total solids concentrations, in 422 calves up to one week of age on 119 southern Ontario dairy farms indicated that 39.8% of calves showed failure of passive transfer (Leslie and Todd, 2007).

Nutritional scour is common in preweaned animals and occurs where calves that are sucking cows ingest excess milk or animals in automated feeding systems are offered large quantities of milk. Irrespective of the cause, invasion by pathogens commonly follows episodes of nutritional scour. The main source of these pathogens is manure. Rotavirus, coronavirus and cryptosporidia are commonly shed in the faeces of healthy cattle. Calves may be subclinically infected by low doses of infection but heavy challenge will cause clinical disease. Clinically infected animals shed vast numbers of pathogens, e.g. coccidia – 80,000 oocysts/g of faeces, *Salmonella* species – 10^8/g of faeces. This leads to a build up of pathogens in the environment leading to more clinical disease. The end result is an outbreak of disease or an increasing incidence of disease during a calving season.

Looking at management factors in the NADIS study it was found that diarrhoea was 3.2 times more likely to occur when calves were reared in groups, 1.9 times more likely when wet bedding was present and 0.6 times more likely when there was no disinfection between groups (Andrews, 2004).

The main source of infection is manure from infected animals. Bio-Mos® has been shown to block the attachment of pathogenic bacteria (*E. coli* and *Salmonella*) to the animal's intestine and colonisation that may result in disease. Penn State University researchers found that Bio-Mos® supplementation (4g/

day) in milk replacer for dairy calves up to 6 weeks of age improved faecal scores and reduced scours (Figures 4 and 5) (Heinrichs *et al.*, 2003).

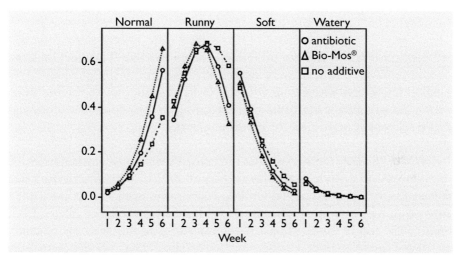

Figure 4. Probability of faecal fluidity scores by week of age for calves fed milk replacer containing antibiotic (Heinrichs et al., 2003).

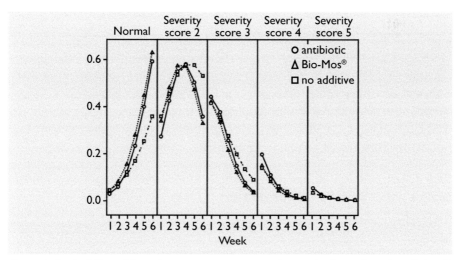

Figure 5. Probability of faecal scours severity scores by week of age for calves fed milk replacer containing antibiotic (Heinrichs et al., 2003).

A recent study (Terré *et al.*, 2007) in calves supplemented with Bio-Mos® (4 g/day), from 10 days of age, showed that the incidence of loose faeces was numerically greater in the control than in Bio-Mos® fed calves (Table 3). All loose faeces occurred within the first 3 weeks, and lasted on average 12.6 and 11.8 days in control and Bio-Mos® calves, respectively.

All medical treatments were performed within the first 3 weeks (Table 4), and no medical treatments were applied after the third week. The number of medical treatments was greater (*P*<0.05) in the control than in the Bio-Mos® calves. The percentage of treated animals during the first 2 weeks of study was 33.3% for control and 13.3% for Bio-Mos® calves.

In the NADIS study (Andrews, 2004) Cryptosporidia was identified in 29.5% of outbreaks. In a study from May to August 2002 *C. parvum* infection was detected in 203 (40.6%) of 500 Ontario dairy calves aged 7 to 21 days in 51 farms with a history of calf diarrhoea. Within farm prevalence ranged from 0% to 70%, and both shedding and intensity of shedding were significantly associated with diarrhoea. Furthermore, calves shedding *C. parvum* were three

Table 3. Incidence (n) and average length (d) of diarrhoea (faecal score greater or equal than 2) in calves supplemented or not with Bio-Mos®.

Treatment	Week 1		Week 2		Week 3	
	n	d	n	d	n	d
Control	9	5.2	10	5.3	6	2.1
Bio-Mos	6	6.2	5	3.8	5	1.8

Table 4. Type, incidence (n), and average length (d) of medical treatment (rehydration) in calves supplemented or not with Bio-Mos®.

Treatment	Week 1		Week 2		Week 3	
	n	d	n	d	n	d
Control	6	16.3	4	2.8	2	2.1
Bio-Mos	3	6.4	1	2	1	3

times more likely to present clinical signs of diarrhoea (Trotz-Williams *et al.*, 2005). In a subsequent study on dairy farms with histories of calf diarrhoea or cryptosporidiosis, faecal samples were collected weekly for 4 weeks from each of 1045 calves under 30 days of age on 11 dairy farms in south-western Ontario during the summer of 2003 and the winter of 2004 (Trotz-Williams *et al.*, 2007a). *C. parvum* oocysts were detected in the faeces of 78% of the 919 calves from which all four faecal samples had been collected. Furthermore, 73% of the 846 calves for which all four faecal consistency scores had been recorded were diarrhoeic at the time of collection of at least one sample. Calves shedding *C. parvum* oocysts had 5.3 times the odds of diarrhoea than non-shedding calves and infected calves shedding more than 2.2×10^5 oocysts per gram of faeces were more likely to scour than infected calves shedding lower numbers of oocysts. Both these studies were carried out in herds with a history of calf diarrhoea or cryptosporidiosis. A recent study carried out during summer and fall 2004 included farms that were experiencing a high frequency of calf diarrhoea as well as farms without neonatal calf diarrhoea problems (Trotz-Williams *et al.*, 2007b). Faecal samples were taken from 1089 calves aged 7–28 days, from 119 herds. Overall, 30% of the calves in the study were shedding *C. parvum* oocysts, with at least one positive calf detected in 77% of herds. Within-herd prevalence ranged from 0 to 80%.

In a recent study (Terre *et al.*, 2007) smear observation to determine presence or absence of *Cryptosporidium* indicated greater (P<0.001) probability of presence of *Cryptosporidium* in faeces of control than in those from Bio-Mos® calves in the first week of the study while no difference was found later on (Figure 6). Generally, *Cryptosporidium* is low the first week of life, and then peaks at second and third week and decreases thereafter. Thus, the evolution found in this study coincides with the normal pattern typically described (Quigley *et al.*, 1994).

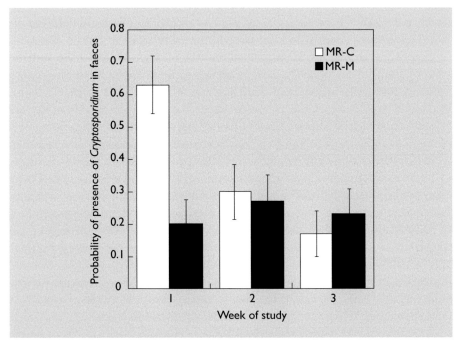

Figure 6. Probability of observing Cryptosporidium spp. in calves feces during the first 3 weeks of the study. Calves supplemented with Bio-Mos® (MR-M) or without (MR-C) in milk replacer throughout the study (Terré et al., 2007).

5. Conclusion

There are three phases of the cycle of calf management – prior to calving, at calving and post calving – and paying attention to and developing procedures for all three will go a long way to ensuring optimum gut health and survival of new born calves.

References

Andrews, A.H., 1999. Calf Mortality. *Cattle Practice* **7**:45-47.
Andrews, A.H., 2000. Calf Enteritis – New Information from NADIS. *UK Vet* **5**:30-34.
Andrews, A.H., 2004. Calf Enteritis – Diarrhoea in the Pre-Weaned Calf – Strategic Investigations of Outbreaks. *Cattle Practice* **12**:109-114.

Biewer, C., 2001. Epidemiologische Erhebungen in einem Praxisgebiet in Franken zu nicht-infektiösen Faktoren mit Einfluß auf Inzidenz und/oder Letalität des akuten Durchfalls junger Kälber (Epidemiological investigations of non-infectious factors with influence on incidence and/or case mortality of acute diarrhoea in young calves in a veterinary practice in Frankonia). Thesis, Ludwig-Maximilians-Universität München.

Brickell, J.S., N. Bourne and D.C. Wathes, 2007. The incidence of calf mortality on dairy farms in southern England. *British Society of Animal Science*, Paper 107.

Carrier, J., S. Godden, J. Fetrow, S. Stewart, P. Rapnicki, M. Endres and P. Mertens, 2005. Studies in dairy cow calving behavior. In: Proceedings Minnesota Dairy Health Conference, St. Paul, MN, pp. 105-110.

Esslemont, R.J. and M.A. Kossaibati, 1996. Incidence of production diseases and other health problems in a group of dairy herds in England. *Veterinary Record* **139**:486-490.

Faber, S.N., N.E. Faber, T.C. McCauley and R.L. Ax, 2005. Case study: effects of colostrum ingestion on lactational performance. *Professional Animal Scientist* **21**:420-425.

Franklin S.T., M.C. Newman, K.E. Newman and K.I. Meek, 2005. Immune Parameters of Dry Cows Fed Mannan Oligosaccharide and Subsequent Transfer of Immunity to Calves. *Journal of Dairy Science* **88**:766-775.

Franklin, S.T., 2006. Building immunity in dairy calves. *Intermountain Nutrition Conference Proceedings*, pp. 21-28.

Girnus, D., 2004. Inzidenz und Verlauf von Neugeborenendurchfall bei Kälbern in einem Praxisgebiet in Oberbayern (Incidence and course of neonatal calf diarrhoea in a veterinary practice in Upper Baveria). Thesis, Ludwig-Maximilians-Universität München.

Godden, S., 2007. Colostrum management for dairy calves. *Proceedings of conference on Calf Management*, Steinkjer, Norway, pp. 7-14.

Heinrichs, A.J., C.M. Jones and B.S. Heinrichs, 2003. Effects of Mannan Oligosaccharide or Antibiotics in Neonatal Diets on Health and Growth of Dairy Calves. *Journal of Dairy Science* **86**:4064-4069.

Heinrichs, A.J. and C.M. Jones, 2003. Feeding of the Newborn Dairy Calf. Penn State College of Agricultural Sciences, Agricultural Research and Cooperative Extension, p. 8.

James, R.E., C.E. Polan and K.A. Cummins, 1981. Influence of Administered Indigenous Microorganisms on Uptake of [Iodine-125] T-Globulin In Vivo by Intestinal Segments of Neonatal Calves. *Journal of Dairy Science* **64**:52-61.

Katikaridis, M., 2000. Epidemiologische Erhebungen zur Kälberdiarrhoe in einem Praxisgebiet in Oberbayern (Epidemiological investigations of neonatal calf diarrhoea in a veterinary practice in Upper Bavaria). Thesis, Ludwig-Maximilians-Universität München.

Leslie, K.E. and C.E. Todd, 2007. Keeping your Calves Healthy. *WCDS Advances in Dairy Technology* **19**:285-300.

McGuirk, S.M. and M. Collins, 2004. Managing the production, storage, and delivery of colostrum. *Veterinary Clinics of North America: Food Animal Practice* **20**:593-603.

Mellor, D.J. and K.J. Stafford, 2004. Animal welfare implications of neonatal mortality and morbidity in farm animals. *The Veterinary Journal* **168**:118-133.

Milk Development Council, 1998. Calf Enteritis & Septicaemia. Project No. 96/R5/10.

Poulsen, K.P., F.A. Hartmann and S.M. McGuirk, 2002. Bacteria in colostrum: impact on calf health. In: *Proc. 20th American College of Internal Veterinary Medicine*, Dallas, TX. (Abstract 52), pp. 773.

Quigley, J.D. 3rd, K.R. Martin, D.A. Bemis, L.N. Potgieter, C.R. Reinemeyer, B.W. Rohrbach, H.H. Dowlen and K.C. Lamar, 1994. Effects of Housing and Colostrum Feeding on the Prevalence of Selected Infectious Organisms in Feces of Jersey Calves. *Journal of Dairy Science* **77**:3124-3131.

Terré, M., M.A. Calvo, C. Adelantado, A. Kocher and A. Bach, 2007. Effects of mannan oligosaccharides on performance and microorganism fecal counts of calves following an enhanced-growth feeding program. *Animal Feed Science and Technology* **137**:115-125.

Trotz-Williams, L.E., B.D. Jarvie, S. Wayne Martin, K.E. Leslie and A.S. Peregrine, 2005. Prevalence of *Cryptosporidium parvum* infection in southwestern Ontario and its association with diarrhea in neonatal dairy calves. *Canadian Veterinary Journal* **46**:349-351.

Trotz-Williams, L.E., S. Wayne Martin, K.E. Leslie, T. Duffield, D.V. Nydam and A.S. Peregrine, 2007a. Calf-level risk factors for neonatal diarrhea and shedding of *Cryptosporidium parvum* in Ontario dairy calves. *Preventive Veterinary Medicine* **82**:12-28.

Trotz-Williams, L.E., S. Wayne Martin, K.E. Leslie, T. Duffield, D.V. Nydam and A.S. Peregrine, 2007b. Association between management practices and within-herd prevalence of *Cryptosporidium parvum* shedding on dairy farms in southern Ontario. *Preventive Veterinary Medicine* (in press).

Veterinary Laboratories Agency, 2006. Cattle disease surveillance. Annual Report: Volume 10. No. 4.

Waltner-Toews, D., S.W. Martin, A.H. Meek and I. McMillan, 1986a. Dairy calf management, morbidity and mortality in Ontario Holstein herds. I. The data. *Preventive Veterinary Medicine* **4**:103-124.

Waltner-Toews, D., S.W. Martin, A.H. Meek and I. McMillan,. 1986b. Dairy calf management, morbidity and mortality in Ontario Holstein herds. II. Age and seasonal patterns. *Preventive Veterinary Medicine* **4:**125-135.

Waltner-Toews, D., S.W. Martin and A.H. Meek, 1986c. The effect of early calfhood health status on survivorship and age at first calving. *Canadian Journal of Veterinary Research* **50:**314-317.

The role of MOS source and structure on antibiotic resistance of pathogenic bacteria

J.M. Tricarico[1] and C.A. Moran[2]
[1]Alltech South Dakota, 700 32nd Avenue South, Bookings, SD 57006, USA
[2]Alltech USA (Corporate Headquarters), 3031 Catnip Hill Pike Nicholasville, KY 40356, USA

1. Introduction

The discovery that antibiotics could be used to control bacterial infections radically changed human and veterinary medicine in the mid 20th century. The application of antibiotic therapy greatly increased the success of medical treatment, prolonged life expectancy and dramatically improved quality of life for both humans and animals. Therefore, antibiotic therapy grew rapidly to control disease in both humans and domesticated animals. The low frequency of spontaneous mutations providing resistance to bacteria led scientists to believe that the development and dissemination of antibiotic resistance would not occur rapidly. However, the extensive exchange of genetic information across bacterial species and unprecedented levels of antibiotic use – resulting from the growth promotion benefits of routinely feeding sub-therapeutic doses – contributed to widely disseminating and perpetuating antibiotic resistance in bacterial populations across a variety of natural environments.

1.1. The origin of antibiotic resistance-encoding genes

Antibiotic resistance-encoding genes are widely believed to have an environmental origin. The most widely accepted hypothesis is that microorganisms developed antibiotics for competitive purposes and antibiotic resistance as the underlying self-protection mechanism (Davies, 1992). The most common antibiotic resistance mechanisms in bacteria are mediated by enzymes and involve inactivation by hydrolysis or inactive derivative formation. Findings of identical tetracycline resistant determinants in mycobacteria and *Streptomyces rimosus* (Pang *et al.*, 1994) and other such similarities in nucleic acid sequences from antibiotic-producing organisms and clinical isolates (Davies, 1992) provide support to this hypothesis and the underlying transfer of resistance genes from producer organisms to pathogens. Nonetheless, there is also evidence supporting that genes implicated in essential enzymatic bacterial cell housekeeping functions, such as kinases and acetyltransferases, may acquire antibiotic-modification abilities through mutation thus conferring antibiotic resistance to bacteria (Salyers *et al.*, 1996). Some researchers suggest that the process of adaptive mutation is an important component of antibiotic resistance generation mechanisms in pathogenic bacteria (Reisenfeld *et al.*, 1997). In this way, a gene encoding an enzyme that inactivates an antibiotic through structural modification can undergo mutations that alter the enzyme's active site and change or expand the spectrum of antibiotic substrates that it can inactivate. The increased frequency at which these mutations appear to occur in response to increased antibiotic selective pressure represents a strong argument for proponents of the adaptive mutation hypothesis.

1.2. Methods for the transfer of antibiotic resistance genes among bacteria

The extensive exchange of genetic material between different species and genera of bacteria is certainly a major contributor to the proliferation of the frequently observed resistance to multiple antibiotics in pathogenic bacteria. The mechanisms for horizontal transfer of genetic resistance determinants between bacteria include transformation, transduction and conjugation. The frequency of both transformation and transduction is not clearly understood in most natural environments and neither are their contributions to the dissemination of antibiotic resistant-encoding genes.

In contrast, conjugation is the most widely studied and appears to be the most important method of genetic transfer between bacteria. The process of conjugation requires cell to cell contact between bacteria and usually involves the transfer of autonomously replicating elements referred to as resistance (R) plasmids that carry the antibiotic resistance genes. Conjugation can also mediate the transfer of other transposable elements (transposons and integrons) that can associate with chromosomes allowing the establishment of antibiotic resistance in bacterial cells that cannot accommodate plasmids. The ability of all these mobile genetic elements to establish themselves in a wide variety of bacteria clearly makes them a critical component to the dissemination of antibiotic resistance in microbial populations. In any case, conjugation requires mating between donor bacteria possessing the mobile element and recipient bacteria lacking the mobile element. Thus, contact between the donor and recipient bacteria is an essential step.

There is also evidence that plasmid transfer increases in nutrient rich environments probably due to greater cell density and increased cell contact (Van Elsas *et al.*, 2000). Therefore, the gastrointestinal tract of intensively reared poultry and livestock provides both a nutrient-rich environment and the constant selective pressure in the presence of sub-therapeutic antibiotic doses that lead to increased development and proliferation of antibiotic resistant bacteria.

1.3. Yeast cell wall preparations as a source of MOS

Yeast call wall mannoproteins from *Saccharomyces cerevisiae* are highly glycosylated polypeptides, often 50-95% carbohydrate by weight, that form radially extending fibriae at the outside of the cell wall (Lipke *et al.*, 1998; Kapteyn *et al.*, 1999). Many mannoproteins carry N-linked glycans with a core structure of Man10-14GlcNAc2-Asn structures very similar to mammalian high mannose N-glycan chains. "Outer chains" present on N-glycans consist of 50-200 additional α-linked mannose units, with a long α-1,6-linked backbone decorated with short α-1,2 and α-1,3-linked side chains. Until recently the identification of proteins in the cell wall has been hampered by the complex nature of the cell wall structure and its relative resistance to simple digestion and extraction. A novel method was developed to tag and identify cell surface proteins using a method based on treating intact cells with a membrane-impermeable biotinylation reagent that specifically reacts with free amino groups (Casanova *et al.*, 1992). Using this method, the identity of approximately 20 cell wall-associated proteins was confirmed (Mrsa *et al.* 1997), although following a genomic approach greater than 40 have been predicted (Smits *et al.* 1999).

Cell wall proteins can be distinguished into two distinct classes, GPI (glycosylphosphatidylinositol) proteins and PIR (proteins with internal repeats) (Kapteyn *et al.*, 1999). The GPI proteins are linked to other cell wall components through a remnant of their GPI anchor and β1,6-glucan cross-links the proteins to β1,3-glucan: an example is the α-agglutinin protein. The PIR are less well understood but in contrast to the GPI proteins, are not posttranslationally modified by addition of a GPI anchor but are highly O-glycosylated (Mrsa *et al.*, 1999). The mannoproteins determine the surface of the yeast cell and are responsible for the cells antigenic behavior. Their extraction in a functionally intact manner is relatively simple on a lab bench scale but has proved to be rather difficult in the larger batch sizes. Regardless, industrial extraction procedures have been developed and are currently used to produce functional mannan oligosaccharides (MOS) extracts from *Saccharomyces cerevisiae* cell wall.

2. Efficacy of yeast cell wall preparations to control the persistence and dissemination of antibiotic resistance in bacteria

The widespread occurrence of antibiotic resistance in potentially pathogenic bacteria clearly represents a hazard for the health and well-being of both humans and domesticated animals. The high costs associated with treatment of persistent infections resulting from antibiotic resistant bacteria in human medicine are undisputed. These high costs along with the need for alternative therapies to control antibiotic resistant bacteria in human disease have raised awareness in the community as a whole. Animal agriculture also faces great economic damage as a result of the widespread occurrence of antibiotic resistant bacteria in modern day poultry, swine and cattle feed operations. The economic impact is not only limited to rising mortality and treatment costs but is also associated to the inefficacy of antibiotics to effectively promote growth and improve animal performance as resistance develops. Finding alternative technologies that may be applied safely and easily in all areas of animal agriculture and that address all of the present issues is extremely important given the current circumstances.

2.1. Bio-Mos® as a growth promoter

Bio-Mos® (Alltech Inc. KY, USA) has been extensively evaluated as an alternative to antibiotic growth promoters. Mannan oligosaccharides have been shown to be effective in improving the health and performance in a variety of species. The use of Bio-Mos® as an alternative to antibiotic growth promoters in calf milk-replacer was first studied in the early 1990's (Newman *et al.*, 1993; Jacques *et al.*, 1994). These investigators observed a reduction in faecal coliforms, an improvement in weight gain and early dry feed intake. In addition, the incidence of respiratory infection was greatly reduced in calves receiving Bio-Mos® supplementation.

These findings sparked a great number of trials in different species to elucidate the mode of action and evaluate the effect of Bio-Mos® as a feed additive. Recently, a series of independent meta-analyses on data gathered from global research studies over a 10-year period have been conducted to investigate the effect of dietary inclusion of Bio-Mos® on the performance of nursery pigs (Pettigrew, 2000); broiler pen trials (Hooge, 2004a) and turkey pen trials (Hooge, 2004b). Meta-analysis is a powerful statistical tool that allows researchers to compare results from a large database of different trials involving a particular product. With each meta-analysis conducted, the reviewers arrived at the same conclusion – Bio-Mos® improves growth performance and should be recommended for inclusion in the respective diet.

The meta-analysis with weaning pigs, involving 55 comparisons (29 separate experiments and 21 research teams) demonstrated a 4.15% improvement in weight gain, 2.34% improvement in feed conversion and 2.08% increased feed consumption (Pettigrew, 2000). In the broiler pen meta-analysis, the antibiotic control and Bio-Mos® diets gave statistically equivalent performance with regard to growth promotion and feed utilisation (Hooge, 2004a). However, Bio-Mos® diets gave -17.2% relative change in mortality averaging by treatment and -18.1% averaging by trial compared to antibiotic control results. This indicated that Bio-Mos® resulted in a significant (P<0.01) improvement on broiler livability compared to the antibiotics evaluated (including avilamycin, bacitracin, bambermycin or virginiamycin at growth promoting concentrations).

Data in turkeys confirm the findings in broilers, with a meta-analysis involving 27 comparisons of turkeys fed Bio-Mos® versus antibiotic-free control diets. The analysis on turkeys demonstrated that on average Bio-Mos® improved body weight (+2.09%, P=0.01) and reduced mortality (P=0.016) whilst tending to lower FCR (-1.47%, P=0.17) relative to the non-medicated control (Hooge, 2004a). The large amount of data available to date clearly establishes that Bio-Mos® is an effective alternative to antibiotics for growth promotion in poultry and livestock.

2.2. Yeast cell wall preparations as a nutritional strategy to control antibiotic resistance bacteria

The issue of reducing the prevalence and dissemination of antibiotic resistance bacteria is certainly more complex and will require a multifactorial approach. It is now clear that removing the selective pressure in the form of the antibiotics is simply one of many strategies that need to be implemented for effectively controlling the propagation and persistence of antibiotic resistance. Yeast cell wall preparations (YCWP) certainly have a potential role as a nutritional technology to control the prevalence and dissemination of antibiotic resistant bacteria within the animal gastrointestinal tract.

An early study (Lou et at., 1995) examined the potential implications of Bio-Mos® on the prevalence of antibiotic resistance and the distribution of tetracycline resistance determinants in faecal bacteria from a swine herd not exposed to antimicrobials for 21 years. Fifty weanling Yorkshire pigs fed a fortified corn-soy base diet were divided into a control and a supplemented group (0.11% Bio-Mos®). Rectal swab samples were collected monthly from study initiation to market weight and 3805 lactose-positive and -negative faecal isolates were tested for resistance to 12 antimicrobial compounds using standardised disk susceptibility tests. Less than 1% of the lactose-positive isolates from each treatment group were resistant to amikacin, chloramphenicol, gentamicin, nalidixic acid, or sulfamethoxazole with trimethoprim. Prevalence of antibiotic resistance in lactose-positive isolates was not influenced by Bio-Mos® treatment.

In contrast, lactose-negative bacteria contained more antimicrobials in their resistance patterns than the lactose-positive bacteria. The proportions of lactose-negative isolates resistant to streptomycin or sulfisoxazole were lower (P<0.01) in Bio-Mos®-supplemented pigs and the proportion of lactose-negative isolates resistant to streptomycin, sulfisoxazole and tetracycline also decreased (P<0.01) in the Bio-Mos®-supplemented group but increased in the control group over time.

These early observations suggested that Bio-Mos® did not increase the prevalence of enteric bacteria resistant to specific antimicrobial agents but rather decreased the prevalence of enteric bacteria harboring multiple resistance in the gastrointestinal tract of pigs not subjected to antibiotic selective forces for a prolonged period of time. Although these early results strongly suggested that Bio-Mos® may be an important component of a comprehensive and effective strategy for decreasing the prevalence of antibiotic resistance in bacterial populations in the gastrointestinal tract of domesticated animals it did not provide any insight into its potential mode of action.

2.3. Yeast cell wall preparations increase the frequency of antibiotic sensitisation in antibiotic resistant bacterial populations

Information encoding antibiotic resistance is most often located in plasmids and extra-chromosomal elements in bacteria. Blocking or reducing the transfer of these plasmids or extra-chromosomal elements is one way to reduce the dissemination and prevalence of antibiotic resistance encoding genes across bacterial populations. Spontaneous elimination of certain plasmids and extra-chromosomal elements is known to occur and represents another approach to controlling the proliferation of antibiotic resistant bacteria. The process of losing antibiotic resistance encoding plasmids in response to the action of certain compounds or conditions is referred to as curing, and those compounds or conditions capable of increasing the frequency of plasmid elimination are known as curing agents (Lakshimi *et al.*, 1989).

Known curing agents include, among others, elevated temperatures, thymine starvation, detergents (e.g. sodium dodecyl sulfate), DNA intercalating dyes (e.g. ethidium bromide), phenolic compounds and certain antibiotics (e.g. rifampin). The modes of action of most curing agents is not well understood but it is clear that most are toxic and may elicit plasmid elimination by interfering with common cellular processes such as DNA replication and processing or maintenance of membrane potential and permeability. Being toxic or hazardous in nature most of these curing agents do not lend themselves to widespread use in animal agriculture but they can be used *in vitro* for comparison purposes to evaluate natural alternatives such as YCWP.

A series of studies were performed at the University of Kentucky with the objective of evaluating the effects of YCWP on the frequency of plasmid elimination and antibiotic sensitisation in *E. coli* and *Salmonella* spp. (Scheuren-Portocarrero, 2004). YCWP increased the sensitivity of *E. coli* to ampicillin, chloramphenicol, streptomycin and neomycin as determined by disk diffusion antibiotic sensitivity testing but did not increase the frequency of plasmid elimination or curing in swine *E. coli* isolates or in a commercial *E. coli* strain containing a broad-host range plasmid.

In contrast, YCWP was capable of curing *Salmonella enteritidis*, *S. monterido*, *S. schwarsengrund* and *S. senf* to levels comparable to ethidium bromide treatment. When curing of *Salmonella* spp. was evaluated over 8 hours incubation, a purified YCWP produced 50 and 65% curing for ampicillin and streptomycin, respectively (Figures 1 and 2). Plasmid losses varied across *Salmonella* spp. with *S. monterido* losing as many as 6 plasmids (Table 1). A purified YCWP was also capable of curing commercial *E. coli* XL1-Blue containing a narrow-host range plasmid.

More recent studies using real-time PCR have confirmed that Bio-Mos® can reduce the expression of genes encoding antibiotic resistance in bacterial populations from the ceca of poultry. Researchers have shown Bio-Mos®-mediated reductions in the mean log copy number of tetracycline resistance genes in caecal contents from broiler chickens (Figure 3; Corrigan and Horgan, 2007a) and turkeys (Figure 4; Corrigan and Horgan, 2007b).

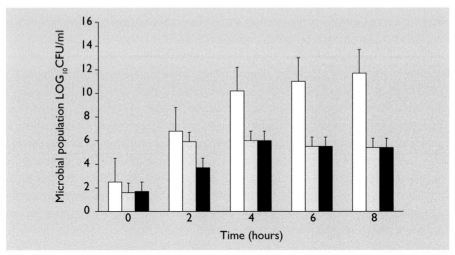

Figure 1. Growth (+SE) of ampicillin resistant Salmonella spp. on medium containing ampicillin (32 µg/ml) following exposure to Bio-Mos® (BM). Bacteria were grown on Muller-Hinton without BM (white), with 0.3% BM (grey) and 0.5% BM (black). Significant (P<0.01) curing was observed at 4 (40%), 6 (50%) and 8 (50%) hours.

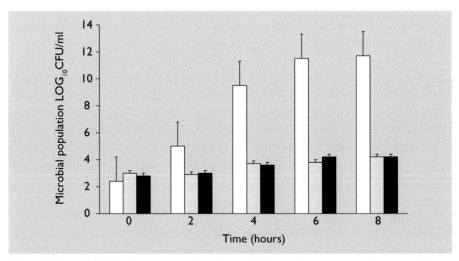

Figure 2. Growth (+SE) of streptomycin resistant Salmonella spp. on medium containing streptomycin (1,000 µg/ml) following exposure to Bio-Mos® (BM). Bacteria were grown on Muller-Hinton without BM (white), with 0.3% BM (grey) and 0.5% BM (black). Significant (P<0.01) curing was observed at 2 (40%) 4 (63%), 6 (65%) and 8 (65%) hours.

Gut efficiency; the key ingredient in ruminant production

Table 1. Selected gram-negative isolates demonstrating plasmid loss or curing following exposure to yeast cell wall preparations (adapted from Scheuren-Portocarrero, 2004).

Strains	# Plasmid lost	Plasmid size (Kb)	Antibiotic Sensitivity Recovered
E. coli XL 1-Blue	1	2.5	Lincomycin and Tetracycline
S. enteritidis	1	2.5	Streptomycin
S. senft	2	2.5, 7	Streptomycin
S. schwartzengrud	1	1	Ampicillin
S. monterido	6	1, 1.5, 2, 2.5, 3.5, 7	Ampicillin, Streptomycin

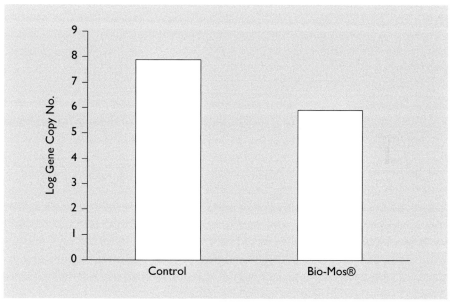

Figure 3. Effects of Bio-Mos® on levels of tetracycline resistance gene tet A in caecal contents of broiler chicks (adapted from Corrigan and Horgan, 2007a). Level of resistant gene tet A was significantly different between control and Bio-Mos® at day 21 (P<0.01).

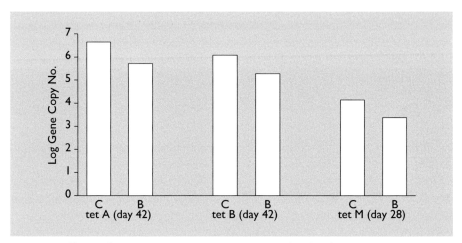

Figure 4. Effects of control (C) and Bio-Mos® (B) on levels of tetracycline resistance genes tet A, tet B and tet M in caecal contents of turkeys (adapted from Corrigan and Horgan, 2007b). Levels of resistant genes were significantly different between control and Bio-Mos® treatments at days 42 (tet A, P<0.050; and tet B, P=0.037) and 28 (tet M, P=0.026).

2.4. Yeast cell wall preparations inhibit the proliferation of antibiotic resistance in bacterial populations

The ability of Bio-Mos® to agglutinate bacteria bearing Type-1 fimbriae is now widely recognised (Spring *et al.*, 2000) underscoring the hypothesis that mannan mediated agglutination may interfere with bacterial conjugation thus reducing the transmission and proliferation of antibiotic resistance among bacteria. A number of studies were performed at the University of Kentucky with the objective of evaluating the ability of yeast cell wall preparations to interfere with bacterial transconjugant formation *in vitro* and in a swine faeces model (Scheuren-Portocarrero, 2004).

In vitro conjugation studies were performed in a nutrient rich broth without antibiotics inoculated with tetracycline-resistant plasmid-bearing donor and ampicillin-resistant plasmid-lacking recipient bacteria. Samples were collected from culture tubes over time and plated onto selective MacConkey agar plates with tetracycline for donor enumeration, with ampicillin for recipient enumeration, and with both tetracycline and ampicillin for transconjugant enumeration. This experimental procedure allowed the researchers to examine, among other factors, the effects of YCWP on transconjugant formation rates.

YCWP reduced (P<0.01) transconjugant formation by 2 log up to the initial 90 minutes of incubation when commercial *E. coli* XL1-Blue was used as donor and *E. coli* MC1000 was used as recipient in this *in vitro* broth model (Figure 5). Supplementation with YCWP every 60 minutes was required to maintain transconjugant formation inhibition over the entire 3 hour incubation period (Figure 6).

YCWP also reduced (P<0.01) transconjugant formation in the swine faeces model although transconjugant formation was generally 1 to 2 log lower in the swine faeces than in the *in vitro* broth model. The doses of YCWP used (0.3 and 0.5%) in these studies effectively reduced transconjugant formation even when the initial recipient concentrations were threefold greater than donor concentrations.

The use of a model analog to Michaelis-Menten and the double-reciprocal Lineweaver-Burk plot to examine transconjugant formation kinetics suggests that crude YCWP inhibits transconjugant formation through a non-competitive mechanism while a purified preparation of YCWP exhibits competitive inhibition (Table 2). Therefore, YCWP clearly reduces transconjugant formation presumably by delaying or blocking the process of conjugation between bacteria. It is apparent that, irrespective of the specific mechanism, reducing conjugation

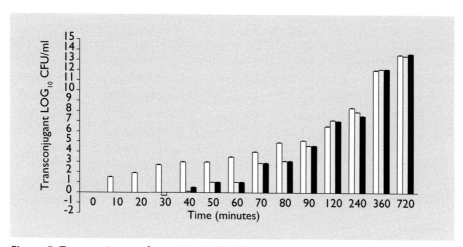

Figure 5. Transconjugant formation (+SD) during in vitro *mating of* E. coli *XL1-Blue (donor) and* E. coli *MC1000 (recipient) cells in the absence (white) and presence of 0.3% YCWP (grey) and 0.5% YCWP (black). Significant (P<0.01) reductions in transconjugant formation were observed at 40, 60, 70, 80 and 90 minutes.*

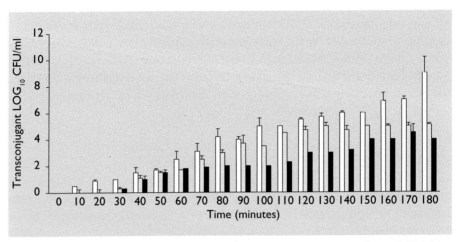

Figure 6. Transconjugant formation (+SE) during in vitro *mating of* E. coli *XL1-Blue (donor) and* E. coli *MC1000 (recipient) cells in the absence of YCWP (white) and following the addition of YCWP (0.5%) at the beginning (grey) and after 60 and 120 minutes of incubation (black). Significant (P<0.01) reductions in transconjugant formation were observed from 50 to 180 minutes.*

Table 2. Conjugation kinetics between E. coli *XL1-blue (donor) and* E. coli *MC1000 (recipient) cells* in vitro *(adapted from Scheuren-Portocarrero, 2004).*

Treatments[1]	Michaelis-Menten Vmax	Linerweaver-Burk Vmax	Type of inhibition	Km
Control	0.42[a]	0.37[a]	No inhibition	8.6×10^{7a}
LB+0.3% P-YCWP	0.18[b]	0.19[b]	Competitive	1.8×10^{7b}
LB+0.5% P-YCWP	0.18[b]	0.2[b]	Competitive	7.3×10^{6c}
LB+0.3% YCWP	0.26[c]	0.31[c]	Non-competitive	2.3×10^{-1d}
LB+0.5% YCWP	0.38[d]	0.39[d]	Non-competitive	5.3×10^{6c}

[1] YCWP = yeast cell wall preparation; P-YCWP = purified yeast cell wall preparation.
[a,b,c,d] Different letter superscripts represent significant differences (P>0.001).

would be a major component in the comprehensive approach to control the dissemination and establishment of antibiotic resistant bacterial populations in the gastrointestinal tract of animals.

Recently, Lee and coworkers (2007) have applied a stochastic compartment modeling for understanding plasmid transmission dynamics and how YCWP could affect plasmid transfer in *Salmonella*. According to their mathematical model, only plasmid transfer rates influence the time required for antibiotic resistance to emerge in a newly introduced population. These researches claim that YCWP may affect plasmid transfer rates by sequestering bacteria in microhabitats that exclude donors.

3. Conclusions

Dietary carbohydrates have traditionally been considered as nothing more than energy yielding molecules and structural components. Recently, studies in the area of glycomics, or the study of oligosaccharides and carbohydrate polymers, demonstrate that non-digestible carbohydrates are central to many biological processes such as cellular metabolism, protein structure and function, cell-to-cell communication and host immunity. Therefore, dietary sources of non-digestible carbohydrates such as YCWP will continue to play important roles in animal production and health and represent a critical nutritional component of the necessarily multi-level approach to control the prevalence and dissemination of antibiotic resistant bacteria in the environment.

References

Casanova, M., J.L. Lopezribot, J.P. Martinez and R. Sentandrea, 1992. Characterization of cell-wall proteins from yeast and mycelial cells of candida-albicans by labeling with biotin – comparison with other techniques. *Infection and Immunity* **60**:4898-4906.

Corrigan, A. and K. Horgan, 2007a. Evaluation of Bio-Mos® performance at reducing the levels of antibiotic resistant bacteria in chicken caecal contents. Poster presented at Alltech's 23rd Annual Symposium, May 20-23, 2007, Lexington, KY.

Corrigan, A. and K. Horgan, 2007b. Evaluation of Bio-Mos® performance at reducing the levels of antibiotic resistant bacteria in turkey caecal contents. Poster presented at Alltech's 23rd Annual Symposium, May 20-23, 2007, Lexington, KY.

Davies, J., 1992. Another look at antibiotic resistance. *Journal of General Microbiology* **138**:1553-1559.

Hooge, D.M., 2004a. Meta-analysis of broiler chicken pen trials evaluating dietary mannan oligosaccharide, 1993-2003. *International Journal of Poultry Science* **3**:163-174.

Hooge, D.M., 2004b. Turkey pen trials with dietary mannan oligosaccharide: meta-analysis, 1993-2003. *International Journal of Poultry Science* **3**:179-188.

Jacques, K.A. and K.E. Newman, 1994. Effect of oligosaccharide supplementation on performance and health of Holstein calves pre- and post-weaning. *Journal of Animal Science* **72**(Suppl 1):295.

Kapteyn, J.C., H. Van Den Ende and F.M. Klis, 1999. The contribution of cell wall proteins to the organization of the yeast cell wall. *Biochimica et Biophysica Acta-General Subjects* **1426**:373-383.

Lakshmi, V.V., S. Padma and H. Polasa, 1989. Loss of plasmid linked antibiotic resistance in *Escherichia coli* on treatment with some phenolic compounds. *FEMS Microbiology Letters* **57**:275-278.

Lee, M., J. Maurer, V. Soni and G. Ewald, 2007. Mathematical modeling for understanding plasmid transmission dynamics and how Bio-Mos® affects plasmid transfer to *Salmonella*. Poster presented at Alltech's 23rd Annual Symposium, May 20-23, 2007, Lexington, KY.

Lipke, P.N. and R. Ovalle, 1998. Cell wall architecture in yeast: New structure and new challenges. *Journal of Bacteriology* **180**:3735-3740.

Lou, R., B. Langlois, K.A. Dawson, G. Cromwell and G. Parker, 1995. Effects of Bio-Mos® on prevalence of antibiotic-resistance fecal bacteria among coliforms of pigs. *Journal of Animal Science* **73**(Suppl.1):175.

Mrsa, V., T. Seisl, M. Gentzsch and W. Tanner, 1997. Specific labelling of cell wall proteins by biotinylation. Identification of four covalently linked O-mannosylated proteins of *Saccharomyces cerevisiae. Yeast* **13**:1145-1154.

Newman, K.E., K.A. Jacques and R.P. Buede, 1993. Effect of mannanoligosaccharide in milk replacer on gain, performance and fecal bacteria of Holstein calves. *Journal of Animal Science* **71**(Suppl 1):271.

Pang, Y., B.A. Brown, V.A. Steingrube, R.J. Wallace, Jr. and M.C. Roberts, 1994. Tetracycline resistance determinants in *Mycobacterium* and *Streptomyces* species. *Antimicrobial Agents and Chemotherapy* **38**:1408-1412.

Pettigrew, J.E., 2000. Mannan oligosaccharides' effects on performance reviewed. *Feedstuffs* **52** (December 25).

Reisenfeld, C., M. Everett, L.J. Piddock and B.G. Hall, 1997. Adaptive mutations produce resistance to ciproflaxin. *Antimicrobial Agents and Chemotherapy* **41**:2059-2060.

Salyers, AA. and N.B. Schoemaker, 1996. Resistance gene transfer in anaerobes: new insights, new problems. *Clinical Infection and Disease* **23**:S36-S43.

Scheuren-Portocarrero, S.M., 2004. Yeast cell wall preparation as a strategy to control antibiotic resistant bacteria *in vitro* and domestic animals. Ph.D. Thesis. University of Kentucky, Lexington, KY.

Smits, G.J., J.C. Kapteyn, H. van den Ende and F.M. Klis, 1999. Cell wall dynamics in yeast. *Current Opinion in Microbiology* **2**:348-352.

Spring, P., C. Wenk, K.A. Dawson and K.E. Newman, 2000. The effects of dietary mannan oligosaccharides on cecal parameters and the concentrations of enteric bacteria in the ceca of Salmonella-challenged broiler chicks. *Poultry Science* **79**:205-211.

Van Elsas, J.D., J. Fry, P. Hirsch and S. Molin, 2000. Ecology of plasmid transfer and spread. In: *The Horizontal Gene Pool* (Ed. Thomas C.M.) Harwood Academic, Amsterdam, NL, pp. 175-199.

Keyword index

A

acetate   12, 23, 38, 39, 65
acid detergent fibre (ADF)   14, 16
acidosis   29, 81
   – acute   15
   – subacute   11, 15
ADF *See:* acid detergent fibre
agglutination   104
ammonia   65, 70
   – detoxification   65
   – plasma concentrations   67
   – ruminal concentrations   63, 65
antibiotic   27, 95, 98
   – growth promoters   98
   – ionophore   27, 47
   – resistance   95, 99, 100, 107
   – resistance encoding genes   95
   – resistance encoding plasmids   100
antibodies   82, 84
   – polyclonal   29
Aspergillus fumigatus   18, 19

B

bacteriophages   28
bicarbonate   13, 25
butyrate   12, 23, 37, 38, 39

C

calving difficulty   83
Campylobacter   21
carbohydrates   11
carbon dioxide   35, 37, 45, 51
chewing time   13, 14
chlorates   28
Clostridium perfringens   18, 19
colostrum   82, 83, 84, 85, 86
competitive inhibition   105
conjugation   96, 105

contamination
   – carcass   22
coronavirus   80, 83
cross-resistance   28
cryptosporidia   86, 88
Cryptosporidium   80, 88, 89

D

degradation rate   62
diarrhoea   80, 86, 89
dietary synchrony   68

E

E. coli   21, 28, 80, 83
   – O157:H7   21, 23, 24, 26, 28
efficiency   13
   – feed   66, 72
   – feed conversion   50
   – microbial   63
Eimeria spp.   80
energy losses   35, 40, 83
enterotoxins   21
essential oils   49

F

faecal shedding   23
failure of passive transfer   86
fat, unsaturated   14
feed
   – costs   61
   – intake   13, 66, 98
   – particles   44
   – physical form   13
feeding level   44
fermentation
   – microbial   61
   – secondary   23
fertility   68

Printed in the United States
by Baker & Taylor Publisher Services